13

U:P:D:A:T:E

Energy resources
for a changing world

John E. Allen

UNIVERSITY OF
LONDON

QUEEN MARY AND WESTFIELD COLLEGE

University of London

CAMBRIDGE
UNIVERSITY PRESS

Update

Update is a unique project in educational publishing. It is aimed at A-level students and first year undergraduates in geography. The objective of the series, which ranges across both physical and human geography, is to combine the study of major issues in the geography syllabus with accounts of especially significant case studies. Each *Update* incorporates a large amount of empirical material presented in easy-to-read tables, maps and diagrams.

Update is produced from the Department of Geography at Queen Mary and Westfield College (QMW) by an editorial board with expertise from across the fields of Geography and Education. The editor is Roger Lee.

We hope that you find the series as exciting to use as we find it to produce. The editor would be delighted to receive any suggestions for further *Updates* or comments on how we could make the series even more useful and exciting.

Preface

The importance of energy resources to human welfare hardly needs justification – some aspect or other of the subject is a part of everyday life, from the personal level (keeping the house warm, paying the electricity bill, filling up the car with fuel) right up to the global (the desperate search for fuelwood in parts of the Third World, political uncertainties in the Middle East, the threat of a warmer climate). As a topic of general concern, it lies at or near the top of the 'green' agenda of the environmentally conscious and it is highly interdisciplinary in nature. Important insights and valuable contributions come from expert practitioners in a wide range of fields – engineers, physicists, chemists, earth scientists, agronomists, economists, politicians, geographers and environmental scientists. Because of its fundamental importance and interdisciplinary nature, the topic of energy resources is to be found within the syllabus of many different A-level subjects and first degree courses.

This volume in the *Update* series has grown out of a course in energy resources developed at QMW over the last 15 years. Although students of environmental science are the intended audience, the course is taken (successfully!) by many others whose principal interest is in biology, chemistry, economics, engineering, European studies, geology and, especially, geography. The approach adopted herein emphasizes the scientific basis of energy resource exploitation but does not ignore the eventual necessity for social and environmental choices to be made within a much broader frame of reference. It should, therefore, prove of use to a wide spectrum of students, encouraging them perhaps to look beyond the boundaries of their particular subject discipline and examination requirements. In particular, in presenting at least some of the issues involved in the debate between the rival claims of fossil fuels, nuclear power and renewable sources of energy, I hope that, while trying to remain impartial myself, I will have stimulated the reader into productive thought and constructive contribution.

I have tried to make the style and layout of the book 'user-friendly'; any suggestions for improvements (within the scope of the *Update* series) would, of course, be welcomed.

Published by the Press Syndicate of the University of Cambridge
The Pitt Building, Trumpington Street, Cambridge CB2 1RP
40 West 20th Street, New York, NY 10011–4211, USA
10 Stamford Road, Oakleigh, Victoria 3166, Australia

First published 1992
Reprinted 1993

Printed in Malta by Interprint Limited

A catalogue record for this book is available from the British Library

Library of Congress cataloguing in publication data applied for

ISBN 0 521 38806 6

The Author

Dr John Allen is Senior Lecturer in Environmental Science in the Department of Geography at QMW.
John's first degree and PhD were in Physics and he was elected a Fellow of the Institute of Physics in 1974.
As a keen environmental scientist he has broadened his range of interests and he now holds degrees in
Geology, Geography and Town Planning. He has taught a wide range of courses in Environmental
Science including a course on energy resources. As well as his academic interests, John Allen has a passion
for music – especially Renaissance music, which he sings, plays and conducts.

Acknowledgements

I am indebted to all the many organizations that provided material for this volume; specific acknow-
ledgements are made at the appropriate places in the text. I would especially like to record my thanks
to the series editor, Roger Lee, for his invitation to write the book and for his guidance, reassurance
and forbearance during its gestation. The figures, tables and overall design are the work of Edward Oliver,
cartographer in the Department of Geography at Queen Mary and Westfield College, University of
London. Last but not least, I gratefully acknowledge the contribution made by Laura Baxter, who took time
from her undergraduate studies (energy resources amongst them!) to enliven these pages with ideas for
cartoons for the encouragement of the initially apprehensive or reluctant reader. The cartoons were drawn
by Gordon Hendry.

Cover
View of experimental wind turbines at Carmarthen Bay, Wales

Contents

Page

1 Introduction   1
   Energy use – then and now
   About this book

2 The use of energy   4
   Types, amounts and flows of energy
   Natural and industrial flows of energy

3 Fossil hydrocarbon fuels   9
   Coal
   Oil
   Natural gas
   Alternative sources of petroleum

4 Nuclear power   20
   Nuclear fusion
   Nuclear fission reactors
   Uranium resources and nuclear capacity
   Thermal reactors in perspective

5 Renewable sources and their use   33
   Overview of renewable energy sources
   Energy storage
   Hydrogen as a storage medium
   Energy distribution

6 Energy from the Sun   39
   Photothermal applications
   Photoelectric conversion
   Photosynthesis and biomass

7 Energy from the sea   51
   Ocean thermal energy conversion (OTEC)
   Power from the waves
   Tidal power

8 Renewable energy on land   59
   Hydropower
   Power from the wind
   Geothermal energy

9 The fifth fuel – energy conservation   70
   The domestic sector
   Industry and transport
   Combined heat and power (CHP)
   In the longer term: essential electricity

10 Resources for a changing world   79

   Appendices   82

   Glossary   89

   Bibliography   91

1 Introduction

- **Energy is necessary for daily survival. Future development crucially depends upon its long-term availability in increasing quantities that are dependable, safe, and environmentally sound. At present no single source or mix of sources is at hand to meet this future need.**
WCED (World Commission on Environment and Development), 1987, p. 168

Energy use – then and now

No energy – no life! Fortunately for humankind, the Sun provides warmth and light on a time scale that is effectively infinite in human terms: its energy is more than sufficient to fuel the circulations of the atmosphere and oceans, and to allow plant life – the basis of the food chain for all other living organisms – to flourish. Food provides people with muscle power for essential daily tasks but life at anything more than a subsistence level requires the utilization of supplementary sources of energy and power. Such energy dependence has tended to increase exponentially with time (fig. 1.1). The use of fire by the first tool-making hunter-gatherers probably raised their daily per capita energy requirements to a level two to three times that of their daily intake of food. By the time that settled agriculture had become established in the Middle East around the fifth millenium BC, the use of animal power had resulted in a further increase by a similar factor. Another doubling of per capita energy use had occurred by the Middle Ages with progressive agricultural developments and an increasing reliance on wind and water power.

Requirements in the more developed countries (MDCs) today are an order of magnitude (i.e.

about ten times) higher still: unlike the modest technological advances achieved before the Industrial Revolution, which were based largely on renewable sources of energy, today's advanced societies consume large and still increasing quantities of the non-renewable fossil fuels coal,

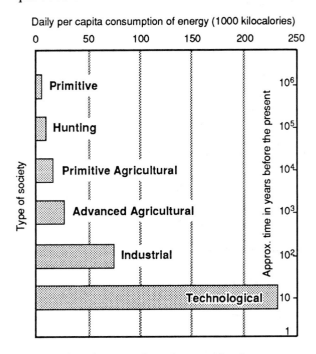

Figure 1.1 **Consumption of energy in the development of human society.**
Source: Cook (1971)

Table 1.1   World energy consumption (million tonnes of oil equivalent Mtoe)

Organisation for Economic Co-operation + Development (1961)

	Coal			Oil			Gas			Nuclear		
	1970	1980	1990	1970	1980	1990	1970	1980	1990	1970	1980	1990
OECD	718	748	883	1623	1818	1729	675	773	812	18	146	365
former USSR & E. Europe	531	564	548	317	562	540	192	381	640	1	21	61
Rest of the world	386	504	801	342	644	828	62	132	255	1	5	24
Total	1635	1816	2232	2282	3024	3097	929	1286	1707	20	172	450

	Hydropower			Total			Population (millions)		
	1970	1980	1990	1970	1980	1990	1970	1980	1990
OECD	222	266	263	3256	3751	4052	719	779	830
former USSR & E. Europe	35	60	73	1076	1588	1862	369	400	428
Rest of the world	48	106	191	839	1391	2099	2577	3203	3948
Total	305	432	527	5171	6730	8013	3665	4382	5206

Source: BP, World Bank (1971, 1981, 1991)

oil and gas. About one-quarter of the world's population lives in the MDCs and consumes three-quarters of the world's production of energy (table 1.1). The maintenance of such energy-intensive economies and their replication amongst the three-quarters of the world's people living in the less developed countries (LDCs) pose two profound questions about the future of world development. Are there sufficient energy resources available to sustain energy-intensive development? What would be the environmental consequences of such high levels of energy consumption?

A consideration of the first of these questions has led many analysts to foresee the development of a widening 'energy gap' in the coming years necessitating immediate increases in supply (fig. 1.2). A contrary view would suggest that the further development of energy supplies to meet increasing demand is just the reverse of the approach required to be consistent with the long-term sustainability of life-support systems on the Earth. Similarly, while many environmentalists see increasing demands for energy as adding to unacceptably intense pressures, including pollution, habitat destruction and climatic change, on an already over-stressed environment, others contend that any such effects are relatively insignificant and localized, and are a small price to pay for vital advances in the human condition.

While sound scientific and technical knowledge is needed to address these issues, solutions are unlikely to be found merely on the basis of such knowledge. Both economic and social deter-

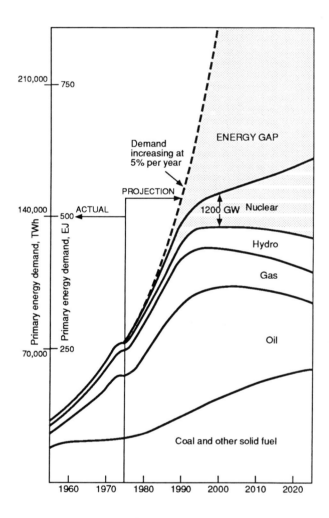

Figure 1.2   The energy gap. This projection, made in 1975, suggests that over the succeeding half-century the possible means of supply of primary energy will fail to match increasing world demand.
Source: RCEP (1976)

minants must be taken into account and value judgements, concerning both objectives and means, must inevitably be made. And even when agreed upon, solutions require political will, legislative instruments and institutional structures for their effective implementation.

About this book

This review aims to provide sufficient information on the nature, availability, applicability and environmental impact of the most important types of energy resource so that the reader may begin both to understand and to be able to contribute intelligently to the energy debate.

In any discussion of energy resources some reminder of the basic rules and limitations of energy technology is needed; this is provided in the next chapter. The bulk of the succeeding text is given over to an examination of the different principal types of energy resource: one chapter each for fossil fuels and nuclear power, followed by three for the considerable variety of renewable sources. Interposed between them comes a brief overview of renewable resource types with some discussion of energy storage and distribution. Chapter 9 considers the important topic of limiting energy demand and the reader is invited finally to consider, with the help of data for a number of contrasting countries, future prospects for the world's energy supply. As well as providing energy data for selected countries, appendices deal briefly with the definition and the classification of resources, and with the measurement and biological effects of ionizing radiation. A selection of possibly unfamiliar scientific terms and concepts are defined in a short glossary.

The questions at the beginning of each chapter (or quotations for the first and last) are intended to draw attention to the principal topics to be discussed. Additional questions at chapter ends are aimed to stimulate further thought about these topics and to broaden the discussion to include some non-technological considerations.

OPEC

Organisation of petroleum exporting countries.

Points to consider

- Comment on trends in levels of energy consumption (gross and per capita), and on changes in the relative proportions obtained from different sources, in both MDCs and LDCs, over the last two decades.

- Can the use of three-quarters of the world's production of energy by one-quarter of its people be justified?

- Do steadily rising standards of living necessarily require the consumption of ever-increasing quantities of energy?

- Suggest (or find out) the nature of the assumptions on which the projection displayed in fig. 1.2 was based. How reasonable were they? What might the projection have looked like if made on the same basis in 1970? Bearing in mind that since 1973 world energy consumption has been increasing at a rate of 2% per annum (compared with 5% during the preceding quarter-century) (Odell, 1989), suggest what factors have influenced this decline and sketch a revised projection for 1990.

- What might be the consequences if the level of energy consumption of the whole world rose to that of its currently most affluent inhabitants? Is this either desirable or practicable?

- As an alternative to trying to fill the energy gap, might it be preferable to try to eliminate it; in other words, should we attempt to match demand with supply instead of vice-versa?

- What, at this stage, do you consider to be the fundamental issues in the energy debate? How can they be resolved most effectively?

¼ World's pop. in MDCs
Consume ¾ worlds prod. of energy
Energy consumption rising in MDCs.
Energy provision low in LDCs.

2 The use of energy

- **What is energy?**

- **How do we measure it – amounts, rates of use?**

- **What rules govern the use of energy and its transformation from one kind to another?**

Types, amounts and flows of energy

In the developed world, we tend to take for granted the continuous availability of unlimited quantities of energy. We use it to help grow and process our food, manufacture our clothes and material possessions, warm or cool our houses, transport us to work or for recreation – the list is a very long one. But in order to compare one type of energy use with another, or the amount used by one nation with another, we need units of measurement.

Energy may be formally defined as the *capacity to do work* – which includes not only the physical work of moving heavy objects but also, for example, the production of heat and electricity, and the chemical changes involved in mental effort by the cells in your brain! The internationally agreed unit for quantities of energy is the *joule*, and one joule (1 J) of energy is required to make a force of one newton (1 N) move through a distance of one metre (1 m). Since an average 0.1 kg apple held in the outstretched hand exerts a downward force of about 1 N, it is seen that the joule is not a large unit – in the space of one hour, a teacher or lecturer might expend one joule of sound energy! For this reason it is often necessary to qualify amounts of energy by the standard SI prefixes, whereby the symbol kJ denotes thousands of joules, MJ millions of joules, GJ thousands of millions (or billions) of joules, and so on (table

Table 2.1   SI prefixes for powers of 10 (orders of magnitude)

Prefix	Symbol	Multiple	Prefix	Symbol	Multiple
deca	da	10	deci	d	10^{-1}
hecto	h	10^2	centi	c	10^{-2}
kilo	k	10^3	milli	m	10^{-3}
mega	M	10^6	micro	μ	10^{-6}
giga	G	10^9	nano	n	10^{-9}
tera	T	10^{12}	pico	p	10^{-12}
peta	P	10^{15}	femto	f	10^{-15}
exa	E	10^{18}	atto	a	10^{-18}

2.1). Thus every tonne of ordinary coal contains about 30 GJ of energy, a typical large modern power station is able to supply 5 TJ of electrical energy every hour and nearly 1 EJ of energy is consumed by the world's population every day.

Energy manifests itself under various guises which can be simply considered as being of two basic types. All moving objects, from bullets to planets, possess *kinetic* energy by virtue of their motion; this category would also include electric currents (the flow of electrons along wires) and the rotational energy of a flywheel, as well the energy of the wind and of rivers. Most other kinds of energy are *potential*, that is stored in some way but available for use. Examples include the gravitational energy of water held behind a dam; the chemical energy of fossil fuels, explosives, batteries and food; and the energy stored in atomic nuclei, from which the Sun derives its huge output and the Earth most of its internal heat. These two basic categories are not, of course, mutually exclusive: for example, both an ocean wave and a cyclist riding downhill have kinetic energy (because of their forward motion) as well as gravitational potential energy (because of their difference in height relative to mean sea level).

In addition to quantifying *amounts* of energy, we also need to be able to specify the *rate* at which it is produced, consumed or transformed from one type to another; this is known as *power*. By analogy with the definition of energy above, power may be formally defined as the *rate of doing work*, the latter term having the same broad meaning as before. The internationally agreed unit of power is the *watt*, equivalent to a rate of energy transformation of one joule per second, that is $1\ \text{W} = 1\ \text{J s}^{-1}$. Thus a typical electric kettle consumes power at a rate of about 2000 W or 2 kW, the power station referred to above is categorized as a 1.5 GW type and (as you could check for yourself) the world's present energy use of almost 1 EJ per day is equivalent to a power consumption of about 10 TW.

Although the standard scientific units for energy and power are joules and watts respectively, for reasons of both historical practice and convenience many other units are in use. The energy content of most of the foodstuffs stocked by your local supermarket, for example, is printed on the outside of the packaging in both kilojoules (kJ) and kilocalories (kcal) – sometimes denoted as Calories (Cal). Your (quarterly) electricity bill assesses your consumption of electrical energy in *units*, which are in fact kilowatt-hours (kWh), while consumption of gas is measured in *therms*, one therm being equal to 100,000 British thermal units (Btu). Workers in the primary energy industries, as well as government departments, often find it convenient to measure amounts of energy in terms of the energy content of fossil fuels, such as million tonnes of coal or oil equivalent (Mtce or Mtoe) or million barrels of oil equivalent (Mboe). Units of power are happily much less diverse than units of energy, the horsepower being the most frequently used alternative. Conversion factors between all of these and other commonly occurring energy and power units are given in table 2.2.

Natural and industrial flows of energy

In both the natural and the human environment energy is constantly being transformed from one type into others. In all such transformations two fundamental laws are rigorously obeyed – the first and second laws of thermodynamics. Although these may be stated in quite formal scientific language, for our purposes it is sufficient to note that the first law, often known as the law of conservation of energy, requires that energy be neither created nor destroyed – it may only be converted from one type into another. Whenever energy appears to have 'gone missing' in any process it is invariably because all losses have not been taken into account: typically these might include radiation, convection and conduction of heat energy to the environment, as well as frictional heat and noise generated by moving parts.

More colloquially, the first law has been expressed as 'there is no such thing as a free lunch' or 'you can't ever get something for nothing'! Put equally informally, the second law tells us that not only do we never get something for nothing, we don't ever break even! In other words, the quality of energy is degraded in transformation and its usefulness for performing work is reduced. The second law is often formally expressed as 'heat does not of itself flow from a colder to a hotter body'; a mathematical formulation of this apparently self-evident truth shows, more usefully, that it is impossible to convert all of a given quantity of heat energy into useful work, as a significant proportion is always discharged to the surroundings.

Table 2.2   Units and conversion factors

Energy

1 calorie (cal) [a]	=	4.1868 J
1 British thermal unit (Btu) [b]	=	1055.06 J
1 kilowatt-hour (kWh) [c]	=	3.60 MJ
1 therm	=	105.506 MJ
1 tonne coal equivalent (tce)	=	29.3 GJ (UN standard)
	=	240 therms (UK – indigenous)
	=	290 therms (UK – imported)
1 tonne oil equivalent (toe)	=	42.6 GJ (UN standard)
	=	430 therms (UK)
1 barrel of crude oil	=	158.98 litres ~ 0.136 tonnes
1 toe	~	1.6 tce ~ 1.1 Gm^3 natural gas
1 electron volt (ev)	=	1.602×10^{-19} J
1 Mev	=	1.602×10^{-13} J (0.1602 pJ)

Power

1 horsepower (hp)	=	745.7 W
1 Btu h^{-1}	=	0.293 W

[a] The calorie, an earlier international scientific unit, was defined as the amount of heat energy required to raise the temperature of 1 g of water (at 15°C) by 1°C.

[b] The equivalent Imperial unit was the British thermal unit, defined as the amount of heat energy required to raise the temperature of 1 lb of water (at 60°F) by 1°F.

[c] This may appear confusing at first (energy measured in power units?) until you realize that kW x h = kJs^{-1} x 3600 s. So 1 kWh is equivalent to 3.6 MJ.

Energy transformations vary considerably in their degree of efficiency but there remains the ever-present tendency for some energy to be 'lost' to the environment as low-grade heat. An ordinary electric light bulb (tungsten-filament) is a familiar household object in which only about 5% of the electrical energy input appears as light, the remaining 95% going to heat up the surroundings. A typical car engine converts about 15% of the energy content of the petrol in the fuel tank into useful mechanical and electrical energy, while a modern coal-fired power station dissipates about two-thirds of the coal's energy to the environment as low-grade heat. The relative efficiencies of a

Table 2.3   Approximate efficiencies of common energy conversion devices and the type of energy transformation involved

Electric generator	95%	mechanical → electrical
Large electric motor	90%	electrical → mechanical
Hydro-electric turbine	90%	gravitational potential → electrical
Domestic boiler	75%	chemical → thermal
Rechargeable battery	70%	chemical ↔ electrical
Fuel cell	60%	chemical → electrical
Steam turbine	45%	thermal → mechanical
Fluorescent light bulb	25%	electrical → radiant
Human body	20%	chemical → mechanical
Solar cell	20%	radiant → electrical
Internal combustion engine	15%	chemical → thermal → mechanical
Steam locomotive	10%	chemical → thermal → mechanical
Incandescent light bulb	5%	electrical → radiant

number of common energy-converting devices are illustrated in table 2.3.

The fundamental process that makes life possible on Earth, the conversion by photosynthesis of solar radiant energy into stored chemical energy, has a maximum theoretical efficiency of almost 30%, although green plants typically achieve annual averages of only 1% or 2%. Nature, bound by the same fundamental laws as people, is not more efficient: the important difference is that nature's energy source is inexhaustible – unlike the fossil fuels on which the developed world largely relies.

So the world has no shortage of energy as such; but locally (and perhaps more widely in the future) people do lack sufficient quantities of those concentrated sources of high-quality energy that we call *fuels*. Primary fuels – like coal, crude oil, natural gas, wood, uranium ore – occur in the environment and can often be used with little or no preparation; secondary fuels – such as coke, petroleum, town gas, charcoal, enriched uranium oxide – are derived from them by processes of conversion and refinement whereby greater convenience is won at the expense of some wastage of energy.

When considering flows of energy through society at a national level it is helpful to make a distinction between primary, delivered and useful energy. Primary energy refers to the heat content of all the sources of energy used in the economy;

although the larger part would normally be accounted for by primary fossil fuels (see above), any contributions from renewable sources, such as hydropower, geothermal or wind power, would be included. Delivered energy is that available to the consumer for direct use in whatever manner is required – cooking, heating, lighting, mechanical work, transport, and all kinds of industrial processes. In this category, electricity is the most significant addition to the secondary fuels listed above.

In a typical developed society, about 30% of the primary energy input is lost in the conversion to delivered energy (fig. 2.1). The useful energy that the consumer gets, in the form of heating, lighting, mechanical work, etc., from the energy delivered is reduced by a further factor of about 30%, due to conversion inefficiencies in the consumer's appliances or processes. A more detailed breakdown of the energy flows in the economy of the UK is depicted in fig. 2.2. There is a clear implication here that a possible solution to impending (or actual) fuel shortages might be found in increasing the efficiency of energy use.

Points to consider

- Identify all the energy transformations involved in getting up from your chair to make yourself a cup of coffee, returning to your studies (or favourite TV programme) and drinking the coffee.

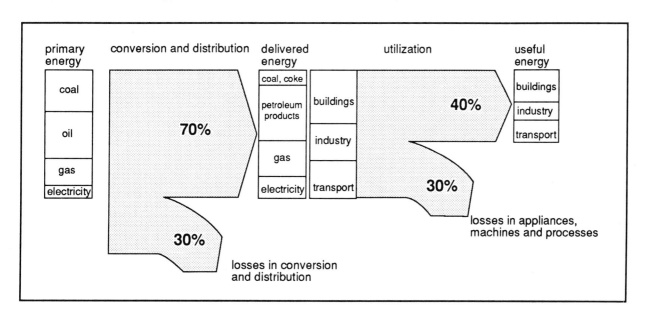

Figure 2.1   Energy flow through a developed country – the UK in 1989. Electricity generated by nuclear power stations and from hydropower and other renewable sources (wind, wave etc.) is conventionally included as primary electricity or energy.
Source: DEn (1990a)

- Your electric kettle, rated at 2.5 kW, should take about $2\frac{1}{2}$ minutes to bring one litre of water to the boil (if you know the relevant equations, check this calculation). In practice, it will take around $3\frac{1}{2}$ minutes; why? Identify the design features of your kettle which are intended to minimize losses of energy to the environment.

- Estimate in kJ (or kcal) your daily intake of food. If all of this energy is ultimately converted to low-grade heat, how does your heat output compare with that of a 100 W electric light bulb?

- In which of your daily activities are you helping to heat up the environment? Explain what could be done about it and justify your response.

- Does the inevitable degradation of high-grade sources of energy (fuels) by use suggest that present patterns of consumption cannot be sustained in the very long term? If not, why not? If so, what sort of changes might be made?

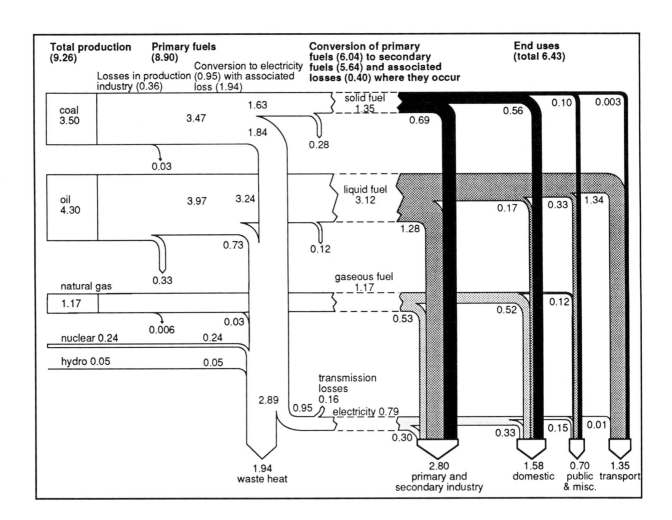

Figure 2.2   **The flow of energy through the UK economy in 1989 (in EJ).**
Source: DEn (1990a)

3 Fossil hydrocarbon fuels

- Where are the world's reserves of fossil fuels located, how large are they and how long will they last?

- What are the advantages and disadvantages of extensive reliance upon coal, oil and gas as sources of energy?

- Are there any practicable alternative sources of hydrocarbon fuels?

Over the past two centuries or so, ways of life in the MDCs have experienced unprecedented change involving marked increases in industrialization, urbanization and affluence. The changes have been made possible by the exploitation of fossil hydrocarbon fuels – coal, oil and gas. Deposits of these resources in commercially exploitable quantities are quite widely distributed around the globe and they provide a very convenient source of concentrated energy for a multitude of applications, especially in the transport sector. Representative values of energy contents are given in table 3.1. Hydrocarbons also have an important subsidiary role to play as sources of organic chemical raw materials in the manufacture of plastics, artificial fibres, pesticides, pharmaceuticals and many other products.

On a human time scale, hydrocarbons are non-renewable and many observers warn that conventional sources of crude oil in particular may be totally depleted before the middle of the twenty-first century.

Carbon dioxide (CO_2), removed by photosynthesis over many millions of years in the distant past, is

Table 3.1 Approximate energy content of hydrocarbon fuels expressed as gross calorific value in MJ kg^{-1}

Peat (dried)		10–15
Coal:	lignite	15–25
	sub-bituminous	20–30
	bituminous	30–35
	anthracite	~33
Coke		25–30
Crude oil		45
Petrol (gasoline)		47
Paraffin (kerosene)		46
Diesel oil		45
Fuel oil		43
Methane (natural gas)		55
LPG (liquified petroleum gas)		50

returned to the atmosphere over a short period of time during the combustion of all types of fossil hydrocarbon fuels. It is estimated that some 6 Gt of CO_2 is released annually from this source (together with smaller contributions from other sources including deforestation). About half this carbon dioxide remains in the atmosphere to constitute the major contribution to the steady increase portrayed in fig. 3.1. It is known that CO_2 is one of a number of gases (others include the natural atmospheric constituents methane, nitrous oxide, ozone and water vapour as well as chlorofluorocarbons (CFCs) of entirely synthetic origin) whose presence in the atmosphere maintains the Earth's average surface temperature some 30°C above its equilibrium value. This is the so-called greenhouse effect and the gases involved are often referred to as 'greenhouse gases'. By using complex but as yet imprecise mathematical models (as well as palaeo-climatic analogues) it has been predicted that increasing concentrations of greenhouse gases will result in significant changes in the Earth's climate during the next century. Moreover it is estimated that CO_2 will be responsible for just over half of the anticipated effects (IPCC, 1990). These are generally agreed to include overall warming by about 3°C (with lesser increases near the equator and greater increases towards the poles) and variations in the amounts and seasonal distribution of precipitation. Such changes would have serious implications both within the hydrosphere (including a general rise in sea level) and the biosphere (especially perhaps for the production of grain in the mid-latitude corn belts). Further discussion of these matters, including the reliability of the predictions and possible strategies for restricting emissions of greenhouse gases or coping with the effects of climatic change may be found in the extensive literature (see, for example, Krause, Bach and Kooney, 1990; Gribbin and Kelly, 1989; Kemp, 1990; Leggett, 1990).

Notwithstanding the inherently finite nature of these resources and their anticipated contribution to any global 'greenhouse' warming, their all-pervasive support for the way of life of the most powerful nations of the world makes it almost inevitable that a major dependence upon them will continue into the foreseeable future. The nature, quantities and location of these resources are therefore of fundamental importance. In any assessment of the availability of a non-renewable resource it is important to distinguish between the total amount conjectured to exist in the world – the *resource(s)* – and that portion of it which is expected to be commercially available with current technology under the prevailing economic climate, the *reserves*. A fuller discussion of this point may be found in appendix 1.

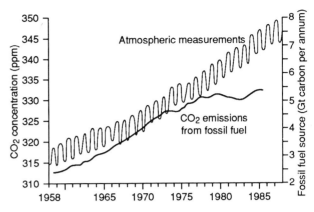

Figure 3.1 The concentration of carbon dioxide (CO_2) in the atmosphere as measured at Mauna Loa, Hawaii, since 1958. Note how, despite an apparent decrease in emissions from fossil fuels, the rise in CO_2 concentrations has continued. One possible explanation attributes this phenomenon to increasing deforestation together with ensuing agricultural practices. Source: Kemp (1990)

Coal

Ordinary domestic coal is perhaps the most familiar form of natural solid fuels, which range from wood at one extreme through peat, whose plant origin is still quite apparent, to anthracite at the other. Peat has a limited use as a fuel. It is relatively easy to collect from its surface deposits, but contains a lot of water, which reduces its energy content or calorific value (table 3.1); it can be considered renewable over time periods of centuries. The transformation of peat-like remains into a higher grade fuel is a very lengthy process requiring the high pressures and elevated temperatures found in the top two kilometres or so of the Earth's surface. The process begins with the build-up of large thicknesses of plant remains, typically in swampy coastal environments including deltas, where a high water table maintains the anaerobic conditions necessary to prevent total decomposition of the organic debris. Bacteria break down the cellulosic components with release of carbon dioxide, water and methane (marsh gas), which results in an increase in the proportion of carbon in the residue. Under the geologically favourable circumstances of a subsiding sedimentary basin, the peaty material will become buried beneath increasing thicknesses of sand, silt and mud.

The effects of elevated temperatures and pressures maintained over millions of years are to decrease steadily the proportion of oxygen, and to some

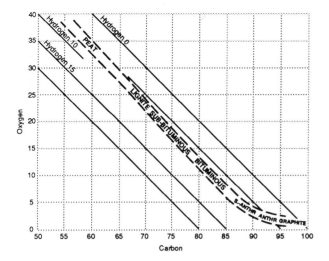

Figure 3.2 **The approximate composition of coals of different rank.** Note that sub-bituminous and lignitous types are also described respectively as hard and soft brown coals.
Source: Raistrick and Marshall (1939)

extent hydrogen, relative to carbon, so raising the calorific value of the fuel (fig. 3.2). The various stages in this process of *coalification* are known as levels of *rank*. Carried to completion, the result would be a deposit of almost pure carbon (graphite). Coals are rarely found younger than about 50 million years (Ma). Important periods for coal formation occurred in the northern hemisphere about 300 Ma ago and about 250 Ma ago in the southern. Before about 400 Ma ago, land plants were at too early a stage of their evolution to provide sufficient remains for coal formation. Although, in general, higher rank is correlated with greater age, it should be borne in mind that geological maturity rather than age is the determining factor. Thus, while anthracites are known from the Canadian Rockies little more than 100 Ma old, some brown coals in the Moscow basin date from more than 300 Ma ago.

In general, the calorific value of a coal increases with rank. Low-rank coals burn freely but with a smoky flame, whereas coals of the highest rank, though difficult to ignite, are almost smoke-free. Most bituminous coals can be converted to coke, another smokeless fuel, by heating in the absence of air. As well as hydrocarbons, coals also contain small but significant amounts of other substances derived partly from the depositional environment and partly from the coalification process. These include clay minerals, carbonates, salt and pyrites which, when the coal is burnt, give rise to unwelcome gaseous emissions and a residual ash. The environmental impact of these emissions and

residues can, however, be reduced to almost any desired low level by a variety of commercially available techniques applicable before, during and after the combustion process.

One of the principal uses for coal, especially in more developed countries, is for the generation of electricity in large power stations (fig. 3.3). A consequence of the combustion process is the production of both sulphur dioxide SO_2 (the amount depending on the sulphur content of the coal) and nitrogen oxides, NO and NO_2 (partly from the inevitable involvement of atmospheric nitrogen). Although the serious local effects of such pollutants in the flue gases have been largely avoided by a policy of dispersal from tall chimneys, a consequence in both Europe and North America has been to increase the acidity of precipitation at a regional scale with deleterious effects on freshwater bodies and vegetation in sensitive localities. The elimination of formerly abundant animal and plant life from many lakes in Appalachia and southern Scandinavia, for example, has been attributed primarily to this cause, while it has been cited as a contributory factor in the die-back of foliage found to be occurring quite widely in coniferous forests (see, for example, Battarbee, *et al*, 1988; ERL, 1983; IEA, 1988; Kemp, 1990; McCormick, 1985; Park, 1987; WCE, 1984b).

It is technically feasible to remove (scrub) the SO_2 from the flue gases by a variety of methods, of which the following two are the most common. The first involves dissolving the gas in a solution of sodium sulphite/bisulphite which is then

Figure 3.3 **The 4 GW coal-fired power station at Drax, near Selby, North Yorkshire.** Note the 259 m tall multi-flue stack designed to facilitate the effective dispersion of pollutants into the atmosphere. Flue gas desulphurization equipment has recently been installed.
Photo: National Power

regenerated as the sulphur dioxide is converted to sulphuric acid or elemental sulphur, both of which have many useful applications. Alternatively, the sulphur dioxide may be combined with limestone to form gypsum, also a useful by-product:

$$2CaCO_3 + 2SO_2 + O_2 \rightarrow 2CaSO_4 + 2CO_2$$

This latter process is the one preferred in the UK; it adds about 10% to the overall costs of electricity generation and will ultimately result in the replacement of about 1Mt of mineral gypsum production. Removal of sulphur dioxide with limestone (as well as a reduction in the emissions of oxides of nitrogen) is both more economically and more effectively carried out within a fluidized combustion bed, but this technology remains at the developmental stage (DEn, 1988).

Particulate material in the flue gases can also be efficiently removed (using bag filters or electrostatic precipitators, for example) and added to the residual pulverized fly ash (PFA), most of which is disposed of by landfill. In the UK, however, up to 40% of this material finds useful commercial outlets in the manufacture of cements, aggregates and construction products.

Coal is most easily obtained by surface ('opencast' or 'strip') mining of shallow deposits although underground methods are more common in countries such as the UK where most of the coal lies in seams deep underground. Although resulting in serious disfigurement of the landscape, the former technique can benefit from economies of scale with a much cheaper and cleaner end product. At the conclusion of operations (and invariably in the UK) it is possible to make extensive restoration to the disturbed landscape. In areas already derelict from an earlier phase of sub-surface mining or associated activities (a not uncommon situation in the UK) restoration may result in a very positive gain (fig. 3.4). Underground methods are necessarily more costly, generally more hazardous for the workforce and liable to disruption from unforeseen geological complexities; they give rise to large volumes of spoil for disposal (as landscaped and vegetated surface tips, for example) and can result in differential subsidence at the ground surface. For these and other reasons, a significant proportion of the coal (typically 30 to 70%) is left in place. The modern highly mechanized coalface tends to produce small grades of coal ideally suited for use in large industrial boilers, especially power stations, rather than for the domestic consumer.

The total resources of coal in the world (> 10,000 Gt), are vast compared with its rate of consumption of almost 5 Gt in 1989. Under present economic conditions and with current technology, some 1000 Gt can be considered as recoverable reserves. The UK has 4.6 Gt of proven recoverable reserves of bituminous coal with perhaps 200 Gt additionally in place, of which 40 Gt is estimated to be recoverable. The world distribution of coal is shown in fig. 3.5 and table 3.2, from which it can be seen that the USA, the former USSR, China and Australia together account for 70% of the total reserves. The low numbers for regions such as Latin America (and the absence of Antarctica) reflect lack of geological knowledge rather than absence of resource.

Because of the size and geographical extent of the resource and the costs of its transport, there is a tendency for regions to approach self-sufficiency; hence the extent of international trade in coal is small in terms of the total amount extracted. The principal exporting countries include Australia, China, South Africa and the USA.

Figure 3.4 An opencast coal mine in operation near Ilkeston, Derbyshire (top), and the restored landscape after mining activity has ceased (below).
Photos: British Coal

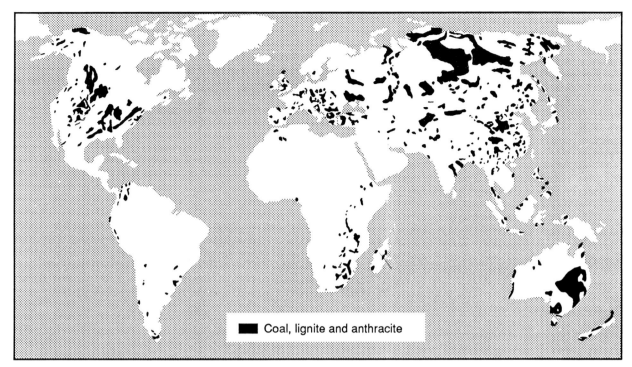

Figure 3.5 **The location of the world's major coal deposits.**
Source: WCE (1981)

Table 3.2 **World proven recoverable reserves of coal**

Area	Proven recoverable reserves of coal (Gt)	% of reserves as semi-bituminous coal and lignite
Latin America	15.6	21.0
USA	215.2	47.5
Canada	7.0	45.0
Western Europe	74.1	61.0
Eastern Europe	92.3	65.6
former USSR	241.0	56.8
China	730.7	16.4
Other Asia	67.9	7.5
Australasia	91.1	50.2
Africa	62.9	0.4

Source: WEC (1989)

Oil

Crude oil is one of a group of naturally occurring hydrocarbons, known collectively as petroleum, that can also be gaseous (e.g. natural gas) or solid (e.g. petroleum jelly). Its composition is dominated by the elements carbon and hydrogen, with smaller amounts of nitrogen, oxygen and sulphur together with trace elements such as vanadium, nickel and chromium. More than 1000 different hydrocarbon compounds have been identified in crude oil; their physical state is largely determined by the number of carbon atoms present. Thus those having 1 to 3 atoms of carbon are gases, 4 to 40 atoms liquids, and more than 40 atoms solids (waxes). The liquid fraction is subdivided into gasoline (petrol) with 4 to 10 carbon atoms, kerosene (paraffin) 11 to 13, diesel 14 to 18, heavy gas oil 19 to 25, and lubricating oil 26 to 40.

Whereas coal is derived from macroscopic terrestrial vegetation, petroleum originates mainly from microscopic marine plant life (phytoplankton), which tends to be most abundant in the shallow, nutrient-rich waters of the continental margins. Sinking after death to the sea-floor, the organic remains may be preserved from immediate decomposition in anoxic black muds. Much like coalification, a long period of burial at appropriately elevated pressures and temperatures is required to generate petroleum, a process known as *maturation* (fig. 3.6). The first *diagenetic* stage of maturation, at relatively low pressures and temperatures (< 50°C), converts the organic remains to *kerogen*, a substance composed of complex organic molecules. These are broken down during the next *catagenetic* stage to simpler hydrocarbons, a process closely akin to cracking in commercial refineries. Maximum conversion to crude oil occurs at slightly higher pressures and

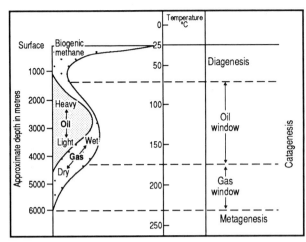

Figure 3.6   The temperatures and approximate depths over which oil and gas are generated from kerogen.
Source: adapted from Brown and Skipsey (1986)

temperatures around 100°C. The precise mix of hydrocarbons in the end product depends mainly on what temperatures were maintained and for how long, as well as the admixture of remains of other organisms (bacteria, zooplankton, land plants, etc.) in the starting material. As with coal, the maturation process (*metagenesis*) will ultimately lead to methane gas and a residue of carbon (fig. 3.7).

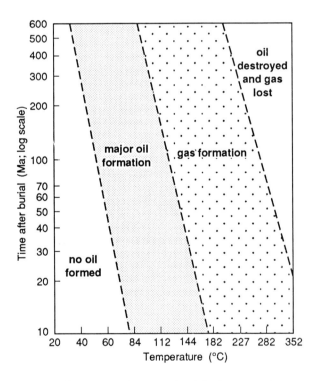

Figure 3.7   A possible relationship between the time after burial of source rocks and the temperature at which oil and gas are formed.
Source: Brown and Skipsey (1986)

Unlike coal, however, which is always extracted from the site where it originated, the only currently useful deposits of oil have migrated from their fine-grained *source rocks* to coarser-grained *reservoir rocks* (such as sandstones or limestones). A seal of an impervious *cap rock* is also necessary to prevent further migration to the surface where all volatile components will be lost. Common types of trap for crude oil are illustrated in fig. 3.8, where, typically, it is accompanied by denser brine and less dense gas, some of which may be dissolved in the oil. In the geological record, the formation of petroleum appears to have been favoured during periods of genial climate and high sea-level when warm, shallow, epicontinental seas were widespread. Many of the known source rocks date from such a period between about 60 and 100 Ma ago. Not surprisingly, less petroleum has survived from similar but more ancient eras.

Petroleum has been located in most of the world's sedimentary basins though infrequently in commercially exploitable quantities. When a reservoir of oil is first tapped, the pressure at depth may be sufficient to drive it to the surface, but normally the oil flowing to the bottom of the well must be pumped out. The flow rate in this stage of *primary recovery* steadily diminishes as the pressure falls and the viscosity of the oil increases leaving typically some 70% of the resource underground. Methods of *secondary recovery* maintain the pressure at depth by pumping down water below the level of the oil or gas above it. Increasing recovery to greater than about 50% involves *enhanced* techniques, such as lowering the oil's viscosity with surfactants or steam, and

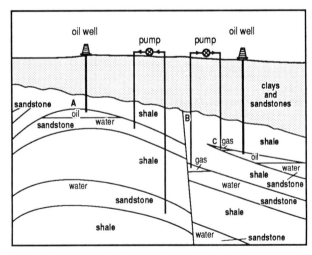

Figure 3.8   Common types of trap for petroleum. Traps include anticlines (A), faults (B) and pinch-outs (C). Possible means of secondary recovery are also shown.

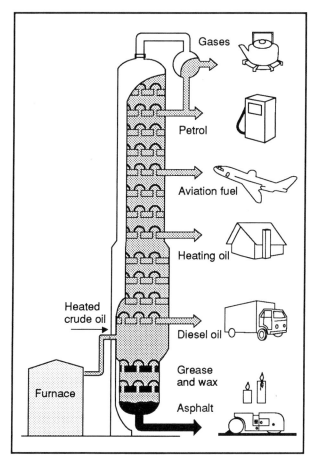

Figure 3.9 **Diagrammatic representation of a distillation tower at an oil refinery showing the principal products, separated according to their boiling points.**
Source: adapted from Miller (1990)

there becomes an increasing risk that the energy expended in recovery of the fuel closely approaches or begins to exceed its energy content. After recovery from a well, crude oil is sent to a refinery where a process of large-scale distillation separates the different fractions according to their volatility (fig. 3.9). A small but important volume of the end products, about 3%, is used as a raw material in the petrochemical industry.

The nature of crude oil and its extraction, transport and refining inevitably places a certain burden on both the workforce and the environment by the risk of spills, explosions and other mishaps. This is especially so since the industry extended its operations from land into an off shore environment. Although oil is a 'cleaner' fuel than coal in respect of particulate pollution, some sulphur dioxide is produced, depending on the sulphur content of the fuel, which inevitably contributes to 'acid rain' effects. The principal end use for liquid petroleum products is as a fuel for all forms of transport. In highly urbanized locations the products of combustion – mainly oxides of carbon and nitrogen together with unburnt hydrocarbons – can give rise to damaging episodes of pollution, especially under conditions of restricted atmospheric circulation. The presence of bright sunlight tends to stimulate the production of photochemical smog containing such undesirable secondary pollutants as ozone and peroxyacetylnitrate (PAN).

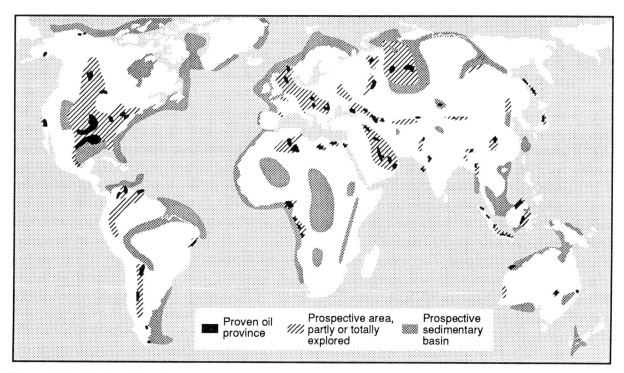

Figure 3.10 The location of the world's principal deposits of petroleum.
Source: WCE (1981)

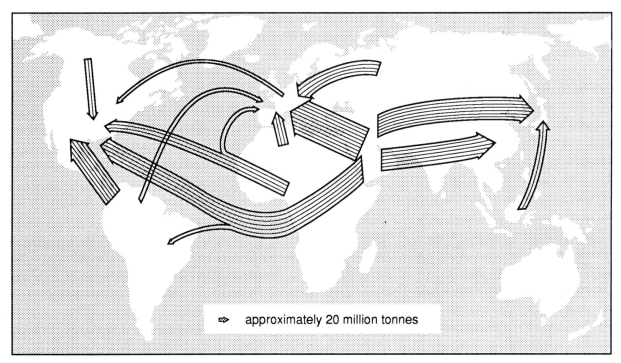

Figure 3.11 The main inter-area trade movements in oil.
Source: BP (1991)

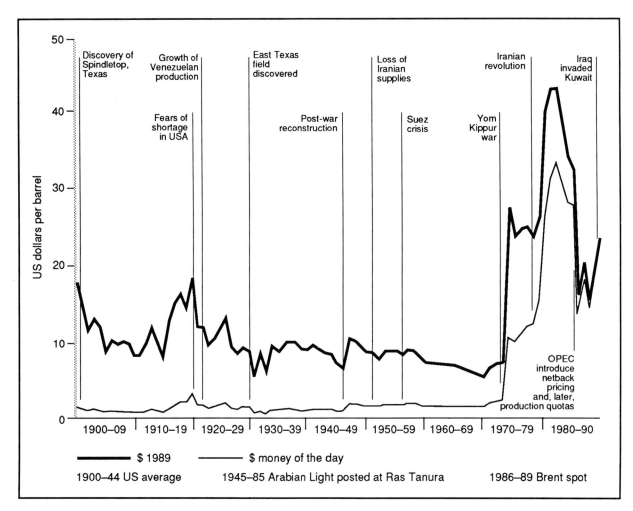

Figure 3.12 Variation since 1900 in the price of crude oil.
Source: BP (1991)

World proven reserves of crude oil are currently estimated to have reached 10^{12} barrels or nearly 140 Gt, a figure which has doubled over the last 25 years, mainly on account of upward revisions in the extent of Middle Eastern deposits. Fig. 3.10 shows that, globally, oil is a very unevenly distributed resource; some 95% of the proven reserves are located in just 20 countries with two-thirds of the total in the Middle East. Reserves in the UK currently stand at 1790 Mt proven, 690 Mt probable and 610 Mt possible (see appendix 1 for a discussion of these terms).

In 1960, a number of the less developed oil exporting countries formed an organization (OPEC) with the aim of influencing the world price of oil so as to encourage its more economical use and to increase revenues. The dominance of world oil supplies and trade (fig. 3.11) by Middle Eastern countries means that events there tend to have very significant and sometimes uncomfortable economic and political repercussions. This can be illustrated by the considerable fluctuations seen in the price of crude oil during this century and especially since 1970 (fig. 3.12).

Consumption of oil is now at a rate of some 65 million barrels daily, the equivalent of about 3 Gt per annum. This would suggest that, at present rates of use, the world has less than 50 years' supply. However, as recently as 1976, the equivalent estimate was only 25 years. Annual consumption in the UK is about 70 Mt for energy generation (a further 20 Mt is used for non-energy purposes and exports). In the more developed world consumption of oil tended to increase during the 1970s but, under the influence of economic recession and greater efficiency, it has declined during the 1980s. In the rest of the world, however, it grew continuously throughout this period with the result that in 1989 world consumption once again reached the peak level recorded in 1979.

Natural gas

As noted earlier, deposits of crude oil are accompanied by a rather 'wet' gas, a mixture of the lightest hydrocarbons. Depending on proximity to markets, as well as technical factors, this gas may simply be 'vented' or 'flared off' for safety, or the less volatile components propane and butane may be removed as liquefied petroleum gas (LPG, and used in rural areas as Calor gas, for example) and the residue, chiefly methane,

cleaned, dried and circulated to consumers as natural gas. It has also been mentioned that methane is produced during the coalification process and many deposits are known (in the southern North Sea, for example) of 'dry' natural gas, quite unassociated with oil (fig. 3.13). Such gas fields are favoured for exploitation since their development does not impede the more vital production of oil. As a fuel, natural gas is considered very 'clean' since, apart from an unavoidable contribution to the atmospheric reservoir of carbon dioxide and possible leakages of methane (a greenhouse gas), it causes no other forms of pollution.

The widespread use of natural gas as a fuel, in part as a replacement for town gas manufactured from coal, is a relatively recent development. As demand has grown over the last 20 to 30 years, so has the effort in exploration, with the result that proven reserves stand now at 113 Tm^3, which, with a calorific value of about 38 MJ m^{-3} or 55 MJ kg^{-1}, is the equivalent of 100 Gt of oil (100 Gtoe). Nearly 40% of these reserves are located in the former USSR and 30% in Middle Eastern countries. Reserves in the UK currently stand at 1265 Gm^3 proven, 625 Gm^3 probable and 585 Gm^3 possible.

Compared with oil, transport of natural gas is difficult; the main international movements are shown in fig. 3.14. Much of this is by pipeline, for example from the former USSR to Europe, from Algeria to Italy and from Canada to the USA. At very low temperatures and additional cost, methane can be converted to liquefied natural gas (LNG) and transported in refrigerated ships as, for example, between Algeria and France, or Indonesia and Japan.

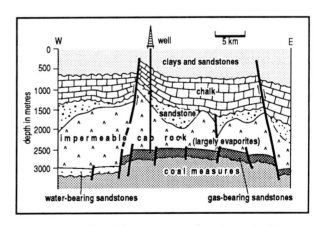

Figure 3.13 Simplified cross-section through the faulted anticlinal gas field of the southern North Sea.
Source: Brown and Skipsey (1986)

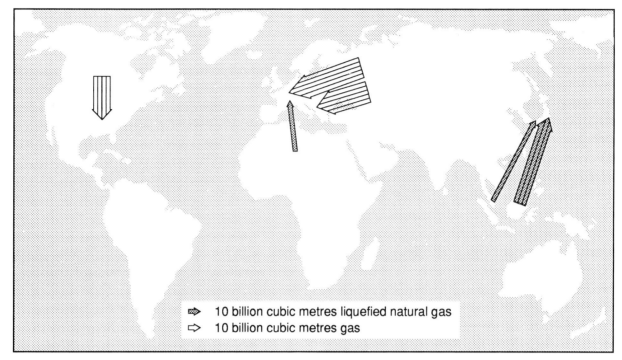

10 billion cubic metres liquefied natural gas
10 billion cubic metres gas

Figure 3.14 The main inter-area trade movements in natural gas.
Source: BP (1991)

World consumption of natural gas shows a consistent upward trend and in 1989 reached the equivalent of 1.7 Gtoe per annum. At present rates of consumption this suggests a lifetime of some 60 years, slightly longer than for oil. Annual consumption in the UK is at a level of about 20 gigatherms (about 55 Gm^3).

Alternative sources of petroleum

The exceptional convenience of liquid and gaseous fossil fuels and our heavy dependence upon them has, especially at times of crisis in their supply, stimulated investigations into possible alternative sources. The more important of these are briefly considered below.

When petroleum migrates through permeable rocks to levels close to the surface it may, through evaporation or the action of meteoric water or bacteria, lose its volatile fractions and produce a sticky deposit of *tar* or *oil sands* (though more correctly described as *bituminous sands*). Less commonly, certain occurrences are thought to represent deposits of normal but immature petroleum. A large deposit, containing some 6 x 10^{11} barrels of asphaltic hydrocarbons, at Athabasca in Alberta, Canada, has been exploited commercially since 1967, producing an average of 45,000 barrels of oil per day. The process of heating the sands to high temperature utilizes only

about 5% of the hydrocarbon resource, and each barrel of oil produced involves the disposal of two tonnes of residual sands, which in turn requires the removal of one tonne of overburden in the opencast mining operation. The upgrading of the bitumen to crude oil gives rise to pollution from sulphur dioxide and heavy metals such as lead and cadmium. A second deposit of significance lies in the Orinoco belt of Venezuela where *pitch lakes* of asphalt have also accumulated in surface depressions, the largest with an estimated reserve of some 6 Mt. Total world proven reserves of bituminous sands are estimated to contain the equivalent of almost 7 Gt of synthetic oil with a further 43 Gt of estimated additional reserves. These are located almost exclusively in Canada, the former USSR and the USA.

Fine-grained sedimentary rocks, occurring at or near the surface, may contain some unaltered kerogen; a familiar example from the English Midlands is provided by the Oxford Clay, extensively used for making bricks which are almost self-firing on account of the high content of organic material. It is technically possible to convert the kerogen content of *oil shales* to crude oil by heating the rock to about 500°C. The 'break even' point, where input of heat energy just equals the calorific value of the oil produced, is at a kerogen content of about $2\frac{1}{2}\%$; a value of 5%, yielding some 25 litres of synthetic crude oil per tonne of shale, is frequently considered to represent an economic lower limit. The Midland Valley of Scotland supported an oil shale extraction industry, yielding about 100 litres per tonne, from 1862 until the suspension of operations in 1963 (DEn, 1975). World resources of such deposits are estimated to contain the equivalent of almost 300 Gt of oil, of which 6 Gt are proven. The largest known deposits occur within the Green River Formation on the borders of Colorado, Utah and Wyoming in the USA, with others in Australia, China, Morocco, Jordan, Sweden and the former USSR. Apart from the energy costs involved, there is also an enormous requirement for water resources in the mining and processing industries, together with the need to dispose of up to two tonnes of shale residue for every barrel of oil produced. *In situ* processing, which involves heating the shale underground in the presence of air and driving the oil so produced to the surface, is technically possible but uneconomic at present.

An obvious starting point for the synthetic production of fluid hydrocarbon fuels, given its abundant reserves, is coal. It should be remembered that coal was used long before oil as a feedstock in the chemical industry and for more than a century in the production of *town (synthesis) gas*, a mixture of carbon monoxide and hydrogen. The technology is now available to convert synthesis gas to *substitute natural gas* (SNG), that is methane, and, either directly or through the intermediate product methanol, to petrol. In fact almost any of the products usually made from crude oil can be derived instead from coal as a starting point. However, the present ready availability of natural gas and crude oil means that this is of only limited practical value except under atypical situations such as that pertaining in South Africa, for example, where coal is unusually cheap and oil and gas depend on historically uncertain imported supplies. In addition to the economic burden of the complex technology, some 30–40% of the energy content of the coal is lost in the conversion process. The underground gasification of coal represents an attempt to reduce some of the economic and environmental costs of *synfuel* (oil and gas substitute derived from coal) production and would be especially appropriate where normal extraction methods are inapplicable (in seams 2 to 3 km deep under the North Sea, for example).

There is considerable potential for further technological developments and lower costs for all of the sources discussed above. (The use of plant and animal material as a starting point for synthetic fuels will be reviewed in chapter 6.)

Points to consider

- What are the most significant environmental impacts attaching to the use of fossil hydrocarbon fuels? What strategies and techniques are available for their amelioration?

- Can you explain why, at the majority of oil wells, the gaseous petroleum fraction is 'flared off' instead of being used as a fuel or chemical feedstock?

- What factors will influence the length of time that the principal fossil fuels are expected to last? Is the 'time to exhaustion' either a meaningful or a useful concept?

- To what extent do you agree with the assertion that although 'it is axiomatic that . . . there must be a limit [to the total stock of fossil hydrocarbon fuels], we neither have the information to say what this is nor any certainty that the substance will still be regarded as a resource when its physical limit is neared'? (Rees, 1985)

- What are the advantages and disadvantages of using synfuels (oil and gas substitutes derived from coal) compared with oil obtained from unconventional sources (oil shales and bituminous sands)?

- To what extent do the problems attached to the use of fossil hydrocarbon fuels outweigh the apparent benefits?

- Is it either necessary or desirable for the more developed nations of the world to be so dependent upon conventional fossil fuels? To what extent did the hostilities of 1990–91 in the Gulf reflect this dependence?

HE'S TRYING TO GET OIL OUT OF SHALE !

4 Nuclear power

- How can useful amounts of energy be obtained from atomic nuclei and in what forms can it be made available?

- What types and quantities of resources are needed and where are they located?

- Is the technology involved cost-effective, reliable, safe and widely applicable?

- What are the benefits and the drawbacks of extensive reliance upon nuclear power as a source of energy?

THE COASTAL SITE IDEA WAS OBVIOUSLY A BAD MOVE!

Without doubt, nuclear power is the most controversial topic in the field of energy studies – an energy source that in more than one sense generates a great deal of heat! Many people are honestly convinced that the harnessing of nuclear power has been the most significant technological development of modern times, one that is capable of solving our present need for increasing supplies of cheap energy from safe, non-polluting sources; others believe with equal sincerity that it represents the gravest evil that the human race has ever inflicted upon itself and should be totally abandoned as soon as practicable. Proponents and opponents maintain with equal conviction that nuclear power can be a long-term or is only a short-term resource; is technologically quite sound or fundamentally flawed; is comparatively cheap or disproportionately costly; causes acceptably low or unacceptably high pollution; is nearly risk free or very hazardous; can be confined to peaceful uses or encourages the spread of destructive weapons; and so on.

This controversy relates in part to the difficulty of comprehending the source of the power and the complexity of the technology required for its controlled release. But it is also sustained by the strong links between nuclear power, military programmes and weapons of enormous destructive capacity, as well as the insidious nature of the radiation hazard and the limited nuclear experience of the human race. For the reader who wishes to pursue the subject of nuclear power more extensively than is possible in this book, there is no lack of literature at all levels (see, for example, Addinall and Ellington, 1982; Blowers and Pepper, 1987; Greenhalgh, 1980; Hunt, 1980; NEA, 1989; Patterson, 1990; Pentreath, 1980; WCE, 1984a).

The constituent protons and neutrons of atomic nuclei are held together by the extremely strong short-range *nuclear force*, which operates over distances of less than about 10^{-15} m. The most tightly bound, and hence the most stable, of all nuclei is iron; as we move away towards either lighter (smaller) or heavier (larger) nuclei, they become progressively less tightly bound (stable). If two nuclei lighter than iron are joined together, or a nucleus much heavier than iron is split asunder into lighter fragments, then the strong nuclear forces have been made to perform work and energy is released. The former process is known as nuclear fusion and the latter as nuclear fission.

Nuclear fusion

The ultimate source of solar energy, and indeed that of all stars, is nuclear fusion and the basic reaction converts hydrogen, the lightest and most abundant element in the universe, into helium. Not surprisingly, perhaps, it has proved impossible so far to reproduce on Earth conditions resembling the solar interior except as an extremely powerful explosion. In order to approach each other sufficiently closely to experience the attractive nuclear force, the hydrogen nuclei (protons) need a kinetic energy equivalent to a temperature of some 15 million°C in order to overcome the electrostatic repulsion engendered by their like positive charges. The process could perhaps be likened to the difficulty of climbing to the top of a very steep hill (corresponding to the electrostatic repulsion) in order to be able to fall into the yawning chasm (of the nuclear attraction) on the other side! The rate of energy release in stars is relatively slow, however, and on Earth temperatures of at least 100 million°C are needed for a useful power source.

The most favourable reaction in terms of probability of occurrence is between the two isotopes of hydrogen, deuterium ($_1H^2$) and tritium ($_1H^3$)

$$_1H^3 + {}_1H^2 \rightarrow {}_2He^4 + {}_0n^1 + \text{energy}$$

with the release of 2.8 pJ (17.6 MeV) as kinetic energy of the two reaction products. Deuterium could be obtained from seawater in effectively unlimited quantities (it is present at about one part in 6500) but tritium, though arising naturally through cosmic ray interactions in the atmosphere, is unstable with a half-life of 12.3 years and could most conveniently be produced from the absorption of neutrons (from the reaction above) by lithium

$$_3Li^6 + {}_0n^1 \rightarrow {}_2He^4 + {}_1H^3 + \text{energy}$$

with the release of 0.77 pJ (4.8 MeV). It is envisaged that a blanket of lithium would surround the fusion reactor vessel. To supply the present annual electricity demand of the UK (almost 300 TWh) would require an initial charge of some 10 kt with a subsequent annual supplement of about 50 t. These requirements seem not unrealistic in view of the known high-grade resources of lithium comprising about 5 Mt. In addition, about 3 t of deuterium would be consumed. A longer-term aim would be to progress from the deuterium–tritium to the deuterium–deuterium fusion reaction, whose resource base would be essentially inexhaustible.

One possible design for a fusion reactor is illustrated in fig. 4.1b. Severe technical difficulties are posed by the need to maintain the reactants, in the form of a very hot plasma, at a high enough density for a sufficient length of time in order that the energy release shall be greater than the energy input (the so-called Lawson criterion). In the design illustrated, known as a tokamak (an acronym in Russian for a toroidal magnetic chamber), containment and energy input rely principally on magnetic fields (fig. 4.1a). A series of vertical coils spaced equally around the doughnut-shaped vacuum chamber create a toroidal field within it. A large plasma current, used for the initial creation and heating of the plasma, is created through the large transformer, of which the plasma forms the single-turn secondary winding. The motion of the large plasma current itself creates a poloidal field around the chamber, and the two magnetic fields interact to form a helical field which effectively confines the plasma, preventing it reaching the chamber walls, except by slow diffusion. Additional poloidal field coils (not shown) placed horizontally above and below the chamber assist with adjustments to the position and shape of the plasma within.

An alternative approach, using the principle of inertial confinement, involves directing very high-powered lasers at milligram sized pellets of reactants. Such a technique allows the necessary high densities and temperatures to be more readily approached but at the expense of shorter confinement times.

The probability of fusion occurring at room temperature is normally exceedingly improbable because of the electrostatic barrier to a sufficiently close approach already referred to. However, the necessary close approach can be facilitated when deuterium and tritium are bound to a muon (an unstable particle with many properties similar to the electron but about 200 times as massive) to form a muonic molecule. This type of muonic

22

fusion has been successfully demonstrated in the laboratory but at present the energy release from fusion reactions is less than the input of energy needed to create the muons. Distortions of the repulsive barrier sufficient to allow fusion to occur can also be realized within a metallic lattice, but the apparent success of such cold fusion reported for deuterium infused into electrodes of palladium has proved difficult to substantiate in subsequent investigations (Bockris, 1991; Close, 1991).

Most research effort at present is concentrated on achieving fusion at very high temperatures. It is envisaged that a commercial reactor of the tokamak variety (fig. 4.1b) would operate in a pulsed mode, each pulse up to one hour in duration. The alpha particles would yield their kinetic energy to the plasma, helping to maintain the necessary high temperature, while the escaping neutrons would be slowed within the surrounding lithium blanket. The heat thus generated would be removed by a circulating coolant and used to generate electricity by means of a conventional steam turbine.

A fusion reactor would be inherently safer than are present-day fission reactors. For example, it would contain only about one minute's worth of fuel (at optimum operating conditions) and could not run out of control. Although no transport or pre-processing of highly radioactive fuels or wastes would be required, the high neutron flux would result in the reactor structure becoming radioactive, which in turn would accelerate the deterioration of its components. Both of these effects, however, could be limited by development of suitable materials and any storage of redundant reactor components limited to decades rather than centuries. The gaseous emissions characteristic of fossil hydrocarbon fuel combustion would not be present.

Over the past few decades the Lawson criterion has been approached increasingly closely: although the required values of the three critical

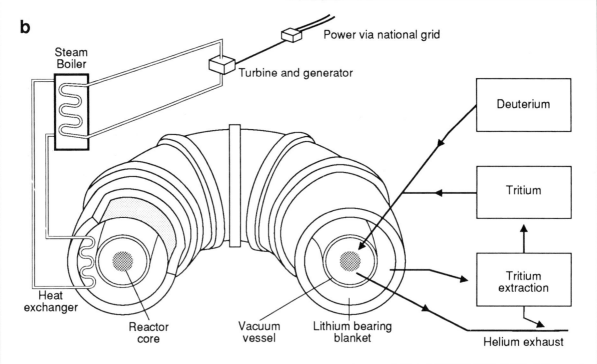

Figure 4.1 (a) The principle of magnetic confinement for thermonuclear plasma employed in the tokamak. (b) Schematic illustration of a possible design for a commercial fusion reactor.
Source: Joint European Torus (JET), Culham Laboratories (promotional literature)

parameters – plasma density, temperature and confinement time – have each been exceeded in particular experiments, their simultaneous achievement, essential for a sustained release of energy, remains an elusive goal. The Joint European Torus at Culham has recently succeeded in sustaining deuterium–tritium thermonuclear reactions for rather less than 20 seconds (Kenward, 1991). The complex, difficult and costly technological problems still to be solved suggest that, without some unexpected breakthrough, fusion reactors could not become a significant source of energy before the second half of the twenty-first century.

Nuclear fission reactors

In nature the spontaneous fission of a heavy unstable nucleus into lighter fragments is a very rare process; on average a uranium atom would wait almost a million, million, million years before its nucleus suffered such a fate. For uranium, alpha-decay is the preferred escape route from instability; since the Earth was formed some 4600 Ma ago, about half the uranium-238 then present has decayed in this way, through a chain of successive alpha- and beta-decays, to the stable isotope lead-206. A second unstable isotope, uranium-235, has a half-life of only 700 Ma and this faster decay rate means that it now makes up only 0.7% of any deposit of natural uranium. Although alpha-decay is accompanied by a release of energy (about 2% of that released by fission), it is far too infrequent a process to be of practical value. Luckily, the uranium-235 isotope has the property that when struck by a slowly moving neutron it absorbs it, forming uranium-236, which then undergoes fission instantaneously. Thus if a sufficient number of these nuclei can be stimulated to split up in this way, useful amounts of energy will be released.

A typical fission reaction might be

$$_{92}U^{236} \rightarrow {}_{54}Xe^{140} + {}_{38}Sr^{93} + {}_{0}n^{1} + {}_{0}n^{1} + {}_{0}n^{1} + energy$$

where the energy released, about 32 pJ (200 MeV), appears principally as kinetic energy of the two fission fragments one of which is usually about half as massive again as the other. These are rapidly brought to rest within the surrounding material and their kinetic energy converted to heat, so raising the ambient temperature. A whole range of different fission fragments having mass numbers between about 50 and 180 may result;

they are themselves unstable and achieve stability mostly through a succession of beta-decays (in the example above to $_{58}Ce^{140}$ and $_{41}Nb^{93}$ respectively). The neutrons (two or three of them may be produced) are very energetic; if they can be slowed down before they escape or are captured by other kinds of nuclei, then it is possible for them to be absorbed by further uranium-235 nuclei, so resulting in yet more fission reactions and the building up of a *chain reaction*.

The neutrons emitted at the moment of fission are known as *prompt neutrons;* in a sufficiently large mass of uranium the progress of the chain reaction is very rapid indeed, resulting in a state of *prompt criticality* and a massive explosion. A controlled release of energy would be virtually impossible were it not for the fact that a small percentage of the unstable fission products also emit neutrons, but after a time interval of seconds or minutes. It is the existence of these *delayed neutrons* that makes commercial nuclear power a practical possibility.

The amount of energy released per fission may seem ridiculously small, but if all the atoms of uranium-235 present in 1 kg of natural uranium were encouraged to release their fission energy together, then some 560 GJ could be instantly available, which is more than ten thousand times the energy content of the equivalent mass of coal. (If you are familiar with Avogadro's number, you could check this calculation.)

The essential features of a commercial nuclear power station, shown schematically in fig. 4.2, can therefore be seen to include:
i) a uranium-based *fuel* (usually uranium dioxide) packaged as a fuel element to allow straightfor-ward replacement and aftercare and in which, for most reactor designs, the proportion of uranium-235 has been enriched from 0.7% to 2% or 3%;
ii) a *moderator* to slow the neutrons down to speeds where they can be absorbed by uranium-235 (water or graphite are most commonly used);
iii) *control rods* to maintain a steady output of energy and prevent a dangerous increase in the ambient temperature (steel impregnated with boron, which has a high affinity for slow neutrons, is often used);
iv) a *coolant* to transport the heat generated away from the reactor core, usually to a heat exchanger, so that high pressure steam may be generated (water is most commonly employed, but UK designs use CO_2);

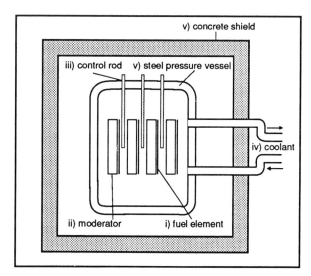

Figure 4.2   Schematic diagram showing the essential features of a thermal nuclear reactor.

v) a steel *pressure vessel* around the reactor core which is enclosed within a *steel-reinforced concrete containment building* for added safety.

Some common reactor designs originating in the UK and USA are illustrated in fig. 4.3; they are known as *thermal reactors*, not because of the heat they generate, but because they utilize slow neutrons which are in thermal equilibrium with their surroundings. The British *advanced gas-cooled reactor* (AGR) has a relatively large core (8.3 m high x 9.1 m diameter) with a low power density (4.5 kW per litre) and uses a relatively low pressure gas coolant (40 bar) to produce steam at 538°C and 160 bar pressure, comparable with a conventional power station. The electrical output is 660 MW per reactor at a thermal efficiency of almost 40%. A typical UK power station comprises two such units with an electrical output of about 1.3 GWe (gigawatts of electricity) for an input of some 3.3 GWth (gigawatts of heat).

The American *pressurized water reactor* (PWR) in contrast has a compact core (3.7 m high x 3.8 m diameter) with correspondingly higher power density (100 kW per litre) and uses a high-pressure water coolant (150 bar), which also acts as the moderator, to produce steam at 290°C and 73 bar. The electrical output is therefore much higher at 1300 MW but the thermal efficiency at 34% is rather lower. Both types of reactor contain about 100 tonnes of fuel. Compared with the AGR, the PWR's compact construction leads to lower capital costs per unit of power generated but the high power density and pressure within the reactor core impose more severe constraints upon safety in respect of overheating.

Throughout the useful life of a reactor, individual fuel elements require replacement every three years or so after about two-thirds of the uranium-235 fuel has been consumed. Some of the products of fission, known as *reactor poisons*, have a marked ability to capture neutrons, so making a chain reaction increasingly difficult to sustain; furthermore, the build-up of fission products generally, together with a degree of radiation damage, produces stress in the fuel elements, so shortening their life.

A proportion of the fast neutrons in the core are absorbed by uranium-238 and the uranium-239 so produced is subject to beta-decay to neptunium-239, which in turn decays to plutonium-239. This latter isotope, like uranium-235, is capable of fission induced by slow neutrons and can therefore be considered a useful fuel. Hence the option arises of recovering the unused uranium and plutonium from the spent fuel elements by *reprocessing*, a treatment of considerable technical difficulty. Whether or not this is carried out, the creation within the fuel elements during their stay in the reactor core of a whole range of unstable isotopes gives rise to serious problems of storage or disposal. As well as generating significant quantities of heat, the decay process involves the emission of biologically hazardous radiation. Hence the fuel elements, or the highly active waste material from their reprocessing, must be cooled for many decades and shielded from human contact for many thousands of years. This whole complex sequence of operations, from the mining of uranium to the disposal of the spent fuel is known as the *fuel cycle* (fig. 4.4).

Elements such as uranium-235 and plutonium-239, which are capable of undergoing fission by slow neutrons and hence are suitable as nuclear fuels in thermal reactors, are known as *fissile materials*. A *fertile material* is one, like uranium-238, that can be transmuted into a fissile material. It is worth noting that the third heavy naturally occurring radioactive element, thorium-232, is fertile, since after absorbing a fast neutron to become thorium-233 it then undergoes two successive beta-decays to produce the fissile isotope uranium-233. It would, however, be costly to use thorium in this way and there are no plans to do so at present.

Thermal power stations have a design life of 25 to 40 years after which time some degree of *decommissioning* is necessary in view of the ever-present radiation hazard. Because of the relative newness of nuclear technology, decommissioning is a procedure of which almost no experience has been gained. After the stripping away of all easily removable material, a possible sequence of events might be a period of *mothballing*, during which the site is kept secure whilst the most intense radiation decays away, followed by permanent *entombment* or total *dismantling*. The operation of decommissioning must tend to add to the volumes of highly active waste material requiring long-term secure storage.

A particular drawback of the thermal nuclear reactor is that it is designed to use only the scarcer uranium-235 isotope, and, as pointed out above, some of this remains unused while a proportion of the plutonium-239 nuclei created do undergo fission. A *fast breeder reactor (FBR)* dispenses with a moderator and uses the *fast* neutrons from the fission reaction to transform the fertile uranium-238 into fissile plutonium-239; this is achieved in a blanket of natural uranium surrounding the core. Reliance for fission on fast neutrons requires a small, highly enriched core. In the British prototype design it is 1 m high by 1.8 m diameter and the uranium oxide fuel is enriched with about 30% plutonium oxide. Metallic sodium, a liquid at moderately high temperatures and low pressures, was chosen as coolant because of its excellent heat-transfer properties and non-moderating effect on fast neutrons. Steam at

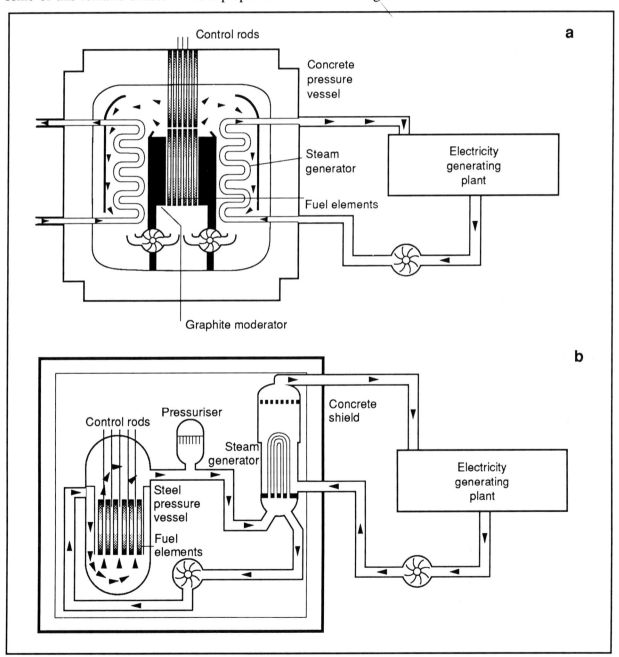

Figure 4.3   Schematic diagram of (a) an advanced gas-cooled reactor (AGR) and (b) a pressurized water reactor (PWR).
Source: Brown and Skipsey (1986)

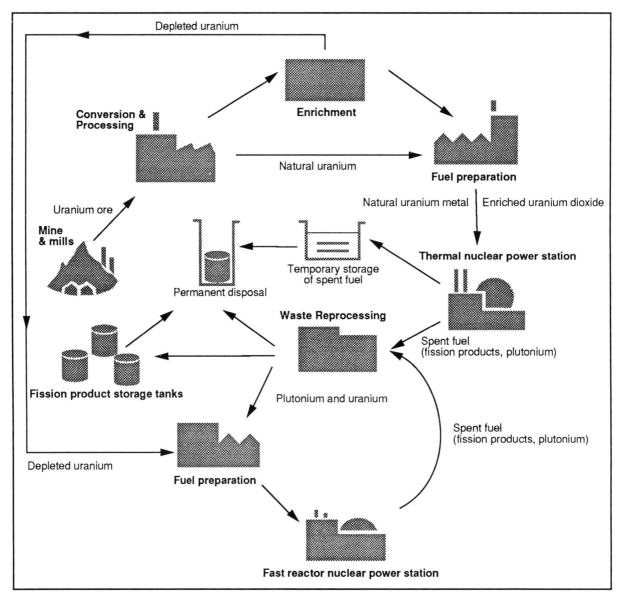

Figure 4.4   The uranium fuel cycle.
Source: adapted from Wild (1980)

590°C is produced which generates 250 MW of electrical power.

In principle a reactor of this type could utilize uranium fuel with about 60 times the effectiveness of a thermal reactor, but the technological requirements are even more demanding and a need for fuel reprocessing is clearly implied. The technology for the breeding of plutonium from uranium and its subsequent separation has its origins in military programmes concerned with the development of nuclear weapons.

Uranium resources and nuclear capacity

The long-term operation of thermal nuclear power stations, and to a lesser degree breeder stations, depends upon a ready availability of their fuel, uranium, which is obtained from a number of different types of mineral deposit, in most of which the uranium is present at a rather low concentration. It is widely disseminated in crustal rocks at a concentration of between 2 and 3 ppm (parts per million) but is not normally considered to be an economic ore deposit below about 350 ppm (0.035%). Uranium may be concentrated by internal igneous and hydrothermal processes, forming primary ore deposits, or recycled by surface processes to form secondary deposits, both types tending to occur within old Precambrian basement or the younger rocks immediately overlying it. The richest deposits identified are primary in nature and together account for about 35% of known reserves. Most of the rest, about 40%, comprise secondary deposits in sandstones which, although of rather low grade, are very extensive. A further 15%, also of low grade, is found in ancient conglomerates.

The anaerobic environment in which black shales, lignite and bituminous coal form, provides ideal conditions for the precipitation of uranium; while a few such deposits contain uranium in commercially valuable concentrations, in others its presence can give rise to a biological hazard in the exploitation and use of the principal resource.

The great diversity in types of uranium deposits naturally results in a similar range of mining techniques. On average, underground methods are three to four times as expensive as open pit methods but this cost differential is often compensated for by a higher grade of ore in the former case. After mining, the ore is crushed in a mill and leached with acid; the uranium content of the solution is extracted by techniques of ion-exchange, then precipitated out and finally dried to form uranium oxide, known as *yellowcake*, which is suitable for shipment. This processing of the ore adds significantly to the total costs and has the effect of reducing the differential between underground and surface mining to little more than a factor of two.

The proportion of uranium resources that can be considered as reserves is quite sensitive to market forces; the considerable fluctuations in the price of uranium since 1968 are depicted in fig. 4.5. The effect of the oil crisis of 1973 was to stimulate interest in nuclear power and to encourage exploration and development of new and often lower-grade sources. In the event, demand for energy grew less than anticipated while disaffection with the nuclear programme increased; overproduction of uranium led to stockpiling and a marked fall in prices. Estimated reserves of uranium are presented in table 4.1. The total of reasonably assured and estimated additional resources at less than $130 kg⁻¹ for the world (excluding the former centrally planned economies) was estimated in 1981 to be some 5 Mt, but by 1986 this had been revised downwards to 3.5 Mt.

The reasons for this revision include an excess of extraction over new discoveries, the transfer of some reserves into higher-cost categories due to rising production costs, the discounting of some of the more speculative of the 'estimated additional resources' category (consequent upon a slight

change in definition), a move away from estimates based on *in situ* quantities to assessments of the amounts recoverable and fluctuations in the exchange rate of the US dollar. (This provides a good illustration of the dynamic nature of the definition of a reserve.)

Little is known of the uranium resources of the (former) centrally planned economies, but, by geological analogy, it has been estimated that the total of assured and speculative resources costing < $130 kg⁻¹ is likely to be 5.2 to 6.5 Mt, the comparable figure for the rest of the world being 13.1 to 15.6 Mt. In this category the UK is estimated to have about 7.5 kt, located as vein deposits in Devon and Cornwall, and disseminated in sandstones in Caithness and Orkney. It is worth bearing in mind that, assuming use only in thermal reactors with no reprocessing, this global figure of some 20 Mt is equivalent in energy terms to perhaps 400 Gt of coal.

Although annual production levels in the rest of the world could be boosted to 70 kt if demand warranted, they remain somewhat below 40 kt. This is sufficient to support almost 250 GW of electrical generating capacity in OECD countries where there is some 50 GW under construction (table 4.2). On average this accounts for close to 25% of electricity requirements but with a wide variation between countries.

Following rapid growth after 1973 (a quadrupling by 1982 and a further doubling to 1988), almost 80% of world nuclear power generation is concentrated in OECD countries. However, a combination of factors, including a slowing in the

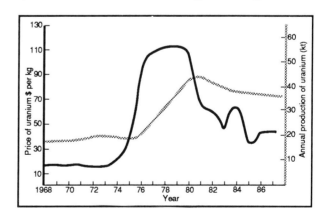

Figure 4.5   Fluctuations in the price and production of uranium between 1968 and 1987. The production data exclude the former centrally planned economies.
Source: adapted from Brown and Skipsey (1986) and WEC (1989)

Table 4.1   **Proven uranium reserves and estimated additional amounts of uranium recoverable (kt) at end 1987**

	Recoverable at less than $US80 kg^{-1}	Recoverable at $US80 kg^{-1} and less than $US130 kg^{-1}
Algeria	26	–
Argentina	9	6
Australia	737	183
Austria	7	10
Brazil	255	–
Canada	258	193
Finland	–	4
France	70	29
Gabon	15	13
Germany (Fed. Rep.)	12	10
Greece	6	–
Greenland	–	43
India	37	25
Indonesia	–	7
Italy	5	1
Japan	–	7
Mexico	5	6
Namibia	124	39
Niger	454	19
Peru	–	3
Portugal	9	1
Somalia	–	10
South Africa	397	139
Spain	35	8
Sweden	3	82
Turkey	–	7
USA	117	270
Zaire	4	–

Estimated total potential resources in (former) centrally planned economies

Bulgaria	50–100
China	1000–2000
Cuba	0–1
Czechoslovakia	100–500
Germany (Dem. Rep.)	100–500
Hungary	10–50
Korea (North)	1–10
Mongolia	10–50
Poland	10–50
Romania	50–100
former USSR	2000–5000
Vietnam	1–10

Source: WEC (1989)

demand for electricity, increased costs and delays in construction of power stations and public disaffection with them following well-publicized accidents at Three Mile Island and Chernobyl, suggests that growth will have declined to about $1\frac{1}{2}\%$ by the end of the century (table 4.2). Larger increases are expected in the former USSR and Eastern Europe.

As fig. 4.6 and table 4.1 show, uranium resources are not at all evenly distributed but concentrated in North America, Australia and Africa. Canada has become the world's leading producer during the last decade due to the exploitation of a rich vein deposit averaging 2% uranium lying beneath 100 metres of glacial overburden at Key Lake, Saskatchewan. A considerable capital investment of $500 million was here richly rewarded. It is generally considered that present sources of supply are adequate to the year 2000, but, thereafter, new sources will need to be developed for which, bearing in mind the typical 15-year lead time between exploration and production, the necessary effort is not being made. This scenario assumes only a modest increase in nuclear power generation which continues to be based upon conventional thermal reactors.

Thermal reactors in perspective

Perhaps the most hotly disputed issue in energy resource studies is the extent to which increasing, or indeed any, reliance should be placed upon nuclear power. The principal limitation of nuclear reactors is that they are a source only of electrical power. They cannot easily contribute to the needs of transport for liquid fuels. Since they do not respond well to fluctuating demands in power output, they are most cost-effectively used to supply the base load of the national electricity supply system; in this respect they are rather less flexible than even the largest coal-fired power stations. In common with the latter, too, they suffer from low thermodynamic efficiencies, discharging large quantities of low-quality heat to the environment. Furthermore, it is more difficult to envisage a useful outlet for much of this waste heat energy, for industrial process heat or as space and water heating for example, in view of the large power rating and remote location of most nuclear installations (chapter 9).

Unlike conventional power stations, however, they do not give rise to large volumes of particulate

Table 4.2 **Actual and propsed nuclear capacity, 1987 (in GWe).** Note that the following OECD countries have no nuclear generating capacity either operational or planned: Australia, Austria, Denmark, Greece, Ireland, Luxembourg, New Zealand, Norway, Portugal, Turkey.

Country	Operational	Under construction	Planned	% nuclear electricity
Belgium	5.5	–	–	66.2
Canada	11.9	3.6	–	15.6
Finland	2.3	–	–	36.4
France	49.4	13.4	2.6	70.0
Germany (Fed. Rep.)	18.9	2.8	2.6	31.2
Italy	1.3	–	–	0.1
Japan	26.3	12.0	14.8	26.1
Netherlands	0.5	–	–	5.2
Spain	6.5	0.9	0.9	31.0
Sweden	9.7	–	–1.3	45.9
Switzerland	2.9	–	–	38.4
UK	9.2	3.1	1.5	18.3
USA	93.6	14.8	–	17.7
Bulgaria	1.8	4.0	2.0	30.0
China	–	2.1	1.2	–
Cuba	–	0.9	2.6	–
Czechoslovakia	3.2	5.1	0.9	25.9
Germany (Dem. Rep.)	1.8	2.6	1.8	9.8
Hungary	1.3	0.4	5.0	18.3
Poland	–	0.9	3.9	–
Romania	–	3.4	0.4	–
former USSR	34.4	–	~80.0	11.2
Yugoslavia	0.7	–	1.0	5.4
Argentina	0.9	0.7	–	11.3
Brazil	0.6	1.2	1.2	0.5
Egypt	–	–	4.0	–
India	1.2	1.4	1.7	2.7
Iran	–	2.4	–	–
Iraq	–	–	0.7	–
Libya	–	–	0.9	–
Mexico	–	1.3	1.0	–
Pakistan	0.1	–	1.8	1.7
S. Africa	1.9	–	–	6.8
S. Korea	5.4	3.7	1.9	53.1
Taiwan	5.1	–	2.0	48.6

Source: IEA (1989c); WEC (1989)

pollutants or noxious gases such as sulphur and nitrogen oxides, nor do they contribute to the atmospheric burden of carbon dioxide and its generally accepted potential to induce climatic warming. These are highly significant advantages and it is in this sense that nuclear power may be said to be 'clean'.

It was frequently maintained in the past that, notwithstanding a much higher capital investment initially, nuclear power stations gave a significant cost advantage over coal-fired types on account of the relative cheapness of the fuel and the lower operating costs. This cost advantage is currently less universally recognized and the differential

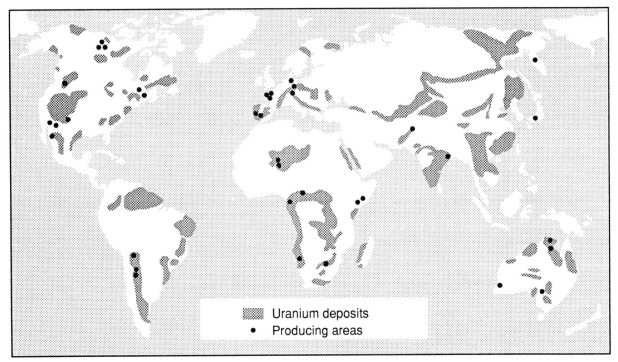

Figure 4.6   The location of the world's principal deposits of uranium ores.
Source: WCE (1981)

varies in any case from country to country. The economics of nuclear power are most favourable where a large scale programme is based on a single proven reactor design with multiple units located at a limited number of sites. Changes in reactor type or design, extended construction schedules, failure to meet safety requirements or performance criteria, and below-forecast capacity factors (the capacity factor is the average power produced expressed as a percentage of the maximum power that could be produced by a reactor or power station) would all have unfavourable effects on costs.

The cost differential is also influenced in most countries by the availability of low-cost supplies of fossil hydrocarbon fuels of appropriate quality, regulations for the control of noxious emissions and, more rarely, by supplies of cheap electricity from renewable sources. It may be further distorted by subsidies or taxes, by assumed discount rates, by the costs of materials and labour, and by failure to extend the appraisal to include the whole fuel cycle. In countries where fuel reprocessing is routinely carried out, such as the UK for example, it is not clear that this is a cost-effective option. Moreover, experience of decommissioning nuclear plant at present relates only to quite small non-commercial reactors; it is considered by some that the costs for large units may rather pessimistically be comparable with costs of construction. A final inevitable

uncertainty relates to future estimates of the relative market prices of uranium and competing fuels.

A further major area of debate concerns the technology of waste management. Most of the waste materials from nuclear power generation are, or have become, radioactive. Thus although relatively small volumes are involved, the disposal of nuclear waste requires special considerations. All living things are exposed to the ionizing radiation which pervades the natural environment and human activities add to this natural background (fig. 4.7). The total annual

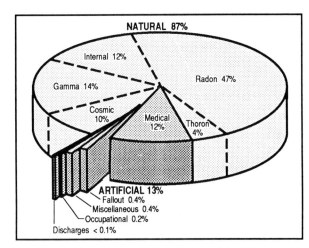

Figure 4.7   The principal sources of ionizing radiation for the population of the UK. The total annual dose is 2.5 mSv.
Source: NRPB (1989)

average dose received in the UK is 2.5 mSv (0.25 rem; see appendix 3 for an explanation of units and the biological effects of radiation) but individual doses will vary depending on lifestyle, diet, medical history, place and type of work or residence. Aircrews, for example, effectively double their natural background dose, while in some indoor environments the annual dose equivalent from the radioactive gas radon can exceed 100 mSv.

While many uses of radiation are undoubtedly beneficial, and some degree of exposure is unavoidable, there is no safe dose – the smallest amount has the potential to cause harm. Clearly, any contribution from nuclear power generation should not increase the risks of harm beyond acceptable limits, which for workers in the nuclear industries could be higher than for members of the general public since the former group experience the risk voluntarily and derive some benefit from it. For nuclear power workers the legal annual limit is set at 50 mSv (the average is in practice about 2 mSv) and for the public 1 mSv (average 0.5 µSv).

On this basis it has been the practice to discharge gaseous and liquid low-level radioactive waste (containing radionuclides with short half-lives) to the environment and bulky solid items to designated shallow landfill. Intermediate-level waste, which includes some long-lived isotopes, tends also to be bulky but to have low heat content. It is incorporated in an inert material such as concrete and is then in a form suitable for dumping at sea (presently under an embargo) or for deeper and more secure landfill (fig. 4.8). In the UK such waste will remain in storage until a suitable site has been developed.

The most intractable disposal problem relates to the high-level waste which, in the UK, refers to the liquid containing all the long-lived actinides and fission products resulting from the reprocessing of spent fuel. Where reprocessing is not practised, then the term refers to the spent fuel itself. In all countries such material is currently stored in cooled tanks because of its considerable heat production. Permanent storage or disposal is envisaged to involve encapsulation of the waste in highly impervious blocks of glass or synthetic silicate rock enclosed within corrosion-resistant containers. After cooling for 50 years or so, their heat output would have decreased sufficiently to permit permanent removal to either a deep stable geological formation on land or possibly beneath

the deep sea floor. Any such repository would need to be absolutely secure for many thousands of years and it is this unpleasant legacy to future generations that makes many people apprehensive.

A final area of doubt surrounds the related problem of reactor safety. Notwithstanding highly complex and sophisticated safety systems and checks, some quite serious accidents involving fuel meltdown have occurred, though very few of them appear to implicate fully operational commercial power plant. These few, however, have received such intense media exposure that scarcely anyone can be unaware of the incidents at Three Mile Island, USA, in 1979 and Chernobyl, Ukraine, in 1986 (WCE, 1989). The latter was of particular significance since it resulted in a spread of radioactivity so far beyond the immediate locality that it could be detected 2000 km away.

While an unauthorized test was being carried out at Chernobyl immediately prior to a scheduled shutdown on the generator of the No. 4 reactor, which involved the disabling of many of the safety systems, the reactor became 'prompt critical' and an extremely rapid build-up of energy (from less

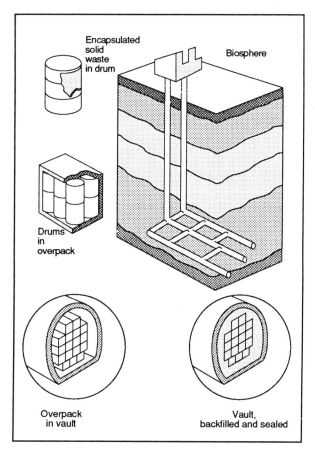

Figure 4.8 **A possible design of a repository for the long-term storage of intermediate level radioactive waste.**
Source: NIREX

than 0.1 to 100 times full power in 4 seconds) caused a massive explosion and subsequent fire. This left the reactor core debris fully exposed and leaking radiation to the environment for a period of 10 days. Within a few days, 31 people had died, nearly two hundred suffered symptoms of acute radiation syndrome and the severity of local contamination obliged the evacuation of 135,000 people from within a 30 km radius. The contamination spread over a large area of Europe and may ultimately result in between 5000 and 100,000 premature deaths from cancer. From the point of view of human health, the fission products of most concern are iodine-131 (with a half-life of 8 days), caesium-134 (2 years), caesium-137 (30 years) and strontium-90 (29 years).

In the UK an estimated 46 μSv was added to the average 50-year burden received; the health effects of this have been estimated as equivalent to no more than 50 additional fatal cancers and would be statistically undetectable. In areas such as the Lake District, where heavy rainfall coincided with the arrival of the plume from Chernobyl, contamination of pastures and consequent restrictions upon sheep farming resulted in some long-term local hardship.

Reactor accidents can generally be attributed to the human factor: the failure of personnel to adhere rigorously to operating rules and procedures, coupled with lack of knowledge and experience in unfamiliar situations, rather than any technical deficiencies in the safety systems themselves. The difficulty of allowing for the unpredictability of human behaviour may cast some doubt on the reliability of risk assessment. Since no energy resource can be exploited entirely free of the risk of casualties, it is ultimately necessary to make a value judgement between the magnitude of the risks involved and the benefits gained.

Points to consider

- What factors might have been responsible for the variations in the market price and annual production figures for uranium seen in fig. 4.5?

- In which countries (or regions of very large countries) would you expect electricity from nuclear power stations to be cheaper than that from coal-fired or other stations and why?

- Bearing in mind the possible options for either 'once-through' fuel cycles or reprocessing, and for thermal or breeder reactors, can you estimate the fuel requirements for an average nuclear reactor over its working life, and hence how large a commitment to nuclear power the uranium mining industry could support and for how long?

- Why might the extent of both past and future commitment to nuclear power (table 4.2) differ so widely between countries?

- Do you regard nuclear power as an appropriate option for extending the energy resource base of LDCs?

- Is it possible to make a meaningful evaluation of the risks associated with the development of different energy resource systems? If it is true, as many experts maintain, that the risks attaching to nuclear power generation are less than for many conventional technologies, why are they unacceptable to a large section of the population?

- Embracing issues wider than those touched on in this chapter (such as the desirability of maintaining a diversity of energy sources, the management and regulation of a nuclear power industry, or the significance of national and strategic interests), do you think that, on the whole, the advantages of nuclear power outweigh its disadvantages?

- A further stage in the development of a possible fusion reactor requires the collaboration of American, European, Japanese and Russian scientists with funding of at least $5 billion (Seneviratne, 1991). On what criteria should the decision whether or not to proceed be based?

5 Renewable sources and their use

- What practicable energy sources are there as alternatives to conventional fossil fuels or nuclear power?

- Where and how can they be exploited?

- How much of our energy needs could they supply, for how long and at what cost?

- Could they help us to manage without nuclear power or fossil fuels altogether?

- In what ways and to what extent does their exploitation affect the environment?

- How necessary is energy storage and what methods are available?

- What is the hydrogen economy?

- How important is the distribution of energy and what are the most practicable methods?

THIS RENEWABLE ENERGY IDEA IS RATHER ENJOYABLE — DO YOU THINK IT WILL CATCH ON ?

Overview of renewable energy sources

Although not strictly applicable to all the energy sources to be discussed, 'renewable' is a convenient label to attach to those which derive from natural flows of energy as opposed to the extraction of non-renewable fossil fuels or uranium. Terrestrial environments are dynamic – ever-changing – and the natural cycles within the atmosphere, hydrosphere, biosphere and at the surface are driven by the most significant source of energy, the Sun. Two further sources make contributions: gravitational energy is involved in ocean tides, for example, whilst the Earth's internal heat powers many geological processes, as is evident from earthquakes and volcanic eruptions.

The relative magnitudes and fates of these natural energy flows are represented in fig. 5.1. It can be seen that, as a group, they constitute a vast resource. The energy received from the Sun in particular is well in excess of the world's present energy consumption but, in general, their rate of delivery is both low and discontinuous. Consequently, the harnessing of renewable sources of energy usually involves relatively high capital investment per unit of energy output, although operating costs are low where the fuel is free. Few of these flows are well suited as base-load or firm power, hence some kind of storage device is necessary for their use in any but a supplementary rôle. Devices that harness renewable sources frequently deliver energy in the form of electricity

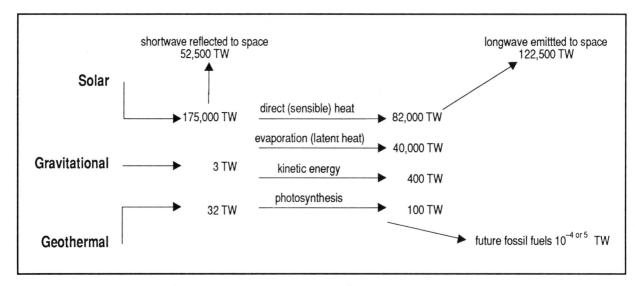

Figure 5.1 Natural flows of energy (in TW) in the terrestrial environment.

and they are generally appropriate for small-scale applications on an incremental basis, as need arises or finances allow. Where a national electricity transmission grid exists, they enable savings to be made in the use of fossil fuels in conventional power stations. The environmental impacts of their exploitation are typically more benign and less extensive than those of fossil fuels and nuclear power (OECD, 1988).

As you study the next three chapters, these simple generalizations may help you identify some of the advantages and disadvantages of the different types of renewable energy, both relative to each other and in comparison with more established energy sources. You may have already realized that an obvious exception to most of the points made above is provided by the traditional and still significant use made, especially in the LDCs, of fuelwood.

For convenience in further discussion, renewable energy sources are divided somewhat arbitrarily into three groups each containing three types: those in the first group derive energy directly from solar radiation (chapter 6), a second group involves those making use of the sea (chapter 7) whilst the remainder are based on land (chapter 8). But first of all, some attention will be given to the important topics of energy storage and distribution.

Energy storage

The desirability of some form of energy store when harnessing most types of renewable energy source in any but supplemental rôles has already been expressed. Consumers in most of the MDCs have

an expectation of power availability at a moment's notice at all times, hence there is an emphasis on 'firm' sources. However, the demand for power is quite variable with well-recognized diurnal and seasonal cycles and it is difficult and inefficient for most firm sources, such as large conventional power stations, to respond rapidly to changes in demand. It is better for complex machinery to operate at a constant level of performance and not be subject to frequent changes in regime; in this way maintenance costs are decreased and lifetimes extended. Energy stores are usually able to respond more rapidly than primary sources to fluctuating demands; they tend also to have limited environmental impacts, with low operating but high capital costs.

Storage can also decrease the amount of firm electrical generating capacity required by helping to meet peak demands; this is the principal function of the pumped storage schemes at Dinorwig and Ffestiniog in Snowdonia (chapter 8). At night, when demand is low, electrical power from large power stations (nuclear or conventional) is taken from the national grid and used by turbines to pump water from a lower to an upper reservoir. At times of sudden increased demand or breakdown of plant, the water is released and, with the turbines switched from pumping to generating mode, electrical power is returned to the grid. As well as providing the most rapid possible response to such an emergency, the system permits large power stations to operate more efficiently. About 1% of electricity generated in the UK is stored in this way; it is retrieved with up to 80% efficiency.

Ideally, because of the inevitable losses incurred

during energy conversions and possible leakages from storage, energy should be stored in the same form in which it will ultimately be required. But – most especially in the case of electricity – this is often not feasible. Most practical storage devices are based upon mechanical, chemical or thermal energy.

Pumped storage provides a large-scale example of a mechanical energy store, in this case gravitational potential energy. Small-scale applications, particularly to transport, have made use of the rotational energy of flywheels. The possibilities of large-scale storage of compressed air at 75 atmospheres in deep salt caverns have been explored in Germany.

Rechargeable accumulators and batteries, such as the lead–acid type used in cars, are a familiar chemical store of electrical energy. For use as a source of power for vehicles they are hampered by a low energy density (about 150 kJ kg^{-1}) when compared with conventional fuels, and a limited useful life. Research and development is aimed at improving both of these characteristics without unduly increasing the capital costs.

Night storage heaters (as used in Economy 7 central heating, for example) are thermal stores of sensible heat. They exemplify many characteristic features of storage devices: although they are long-lasting, technically straightforward, pollution-free and have low maintenance costs, they also have rather high capital costs, are cumbersome and their energy output decreases as the store is depleted. They use refractory bricks as the storage medium but, on a larger scale, water or rocks are most common and they are typically deployed in conjunction with active and passive solar heating systems (chapter 6). At the domestic scale, diurnal fluctuations in heat demand are easily coped with, but inter-seasonal heat stores are more effective on a communal basis since, while the quantity of heat stored is proportional to the volume of the store, heat losses increase only in proportion to the surface area. At the Centre for Alternative Technology, Machynlleth, for example, an interseasonal store of 100,000 litres of water provides heating for the exhibition hall from 100 m^2 of solar panels. A number of large communal heat stores, using either water or rock, are operational in Sweden.

The size of store can be reduced and its output maintained at a more constant level if storage is mainly in the form of latent heat. (A contribution from sensible heat will normally be involved as well.) For example, Glauber's salt ($Na_2SO_4.10H_2O$) crystallizes at 32.4°C with the release of 250 kJ kg^{-1} and has formed the basis of a commercial system. However, cost can be an inhibiting factor and the number of substances which solidify at a reasonable temperature with a sufficiently high latent heat of fusion but which are non-toxic, non-corrosive and do not degrade after repeated temperature cycling is quite limited.

Still higher energy densities can be achieved with thermochemical systems which utilize exothermic reactions, for example the liberation of 3.6 MJ kg^{-1} by anhydrous sodium sulphide when it dissolves in water. The two reactants in separate containers are connected by a valve-controlled evacuated pipe. Opening the valve allows water vapour access to the sodium sulphide, thus beginning the production of heat. The water store must not be allowed to cool too much if the supply of vapour is to be maintained; any low-grade heat source is suitable. The heat output of the 'Tepidus' commercial system is 50 to 60°C above that of the water reservoir. The reaction is reversed when required by driving off the water from the sulphide solution with heat from a higher-grade source such as a solar panel. Again, the difficulties lie in finding suitable compounds that are inexpensive and neither toxic nor corrosive.

Hydrogen as a storage medium

It was noted earlier that many renewable sources of energy are most readily harnessed as producers of electricity, and this limitation applies also to nuclear power. On the other hand, attention has been drawn (chapter 2) to the dependence of the developed world in particular on hydrocarbon fuels, which are virtually essential for transport by air and road. As a possible means of resolving this incompatibility between the forms in which energy can most conveniently be supplied and in which it is required for use, and one that incidentally avoids the most serious pollution problems associated with the combustion of fossil fuels, it has been proposed that hydrogen might ultimately replace petroleum as the principal energy store and fuel. Such a scenario has been dubbed the 'hydrogen economy' (see, for example, Gregory (1973); Crabbe & McBride (1978); Bockris and Veziroglu (1991)).

Hydrogen gas burns readily in air, combining exothermically with oxygen to release 12 MJ m^{-3} or

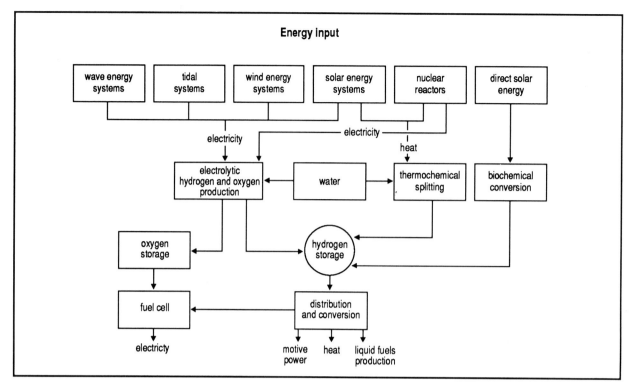

Figure 5.2   Elements of the hydrogen economy.
Source: Brown and Skipsey (1986)

142 MJ kg^{-1}; the obvious drawback is that as a gas it has a low density but, as a liquid, it boils at $-253°C$. It does, however, form stable metallic hydrides from which it can be released by gentle heating, for example

$$Fe\,Ti\,H_{1.7} \rightarrow Fe\,Ti\,H_{0.1} + 0.8\,H_2$$

which occurs at about 50°C. Such compounds, however, are rather costly and rather weighty. The principal advantages of hydrogen as a fuel include ease of transport and pollution-free combustion.

Hydrogen is most conveniently produced by the electrolysis of water. This has been envisaged as a particularly appropriate option for the storage and transport of energy generated as electricity by the marine renewable sources OTEC and wave power (chapter 7), if they should be located at some considerable distance from the shore. Other possibilities for the dissociation of water include thermochemical, which, though normally requiring a temperature of 2500°C, will take place at about 650°C in the presence of a ferrous chloride catalyst; photochemical, involving manganese, ruthenium or titanium complexes having functions analagous to magnesium in chlorophyll; and photobiological, the release of free hydrogen ions by bacteria (e.g. *Clostridium*) and algae (e.g. *Anabaena*).

Apart from energy release by combustion, hydrogen can be converted directly to electricity in a fuel cell, which operates essentially as the reverse of electrolysis. Air (oxygen) is supplied to the cathode, and hydrogen molecules to the anode, where they are dissociated by platinum or gold catalysts. The basic reactions are

$$\tfrac{1}{2}O_2 + H_2O + 2\,e^- \rightarrow 2\,OH^-$$
$$2\,H + 2\,OH^- \rightarrow 2\,H_2O + 2\,e^-$$

whose net effect is

$$2\,H + \tfrac{1}{2}O_2 \rightarrow H_2O$$

with the movement of electrons providing a current through the external circuit connecting the cathode and anode. Cells of 6 kW and $12\tfrac{1}{2}$ kW power output were developed for the Apollo space programme and the Space Shuttle programme respectively, having efficiencies of about 60%, but such devices are as yet rather too costly for general application.

Where a hydrocarbon source is conveniently available, as in organic wastes, coal or low-grade oil (from oil shales or bituminous sands, for example), gasification techniques (chapter 6) may be used to produce methane (substitute natural gas) and liquid fuels such as methanol and ethanol. It would also be possible to derive other non-hydrocarbon liquid fuels including hydrogen peroxide (H_2O_2) and hydrazine (N_2H_4).

The hydrogen economy (fig. 5.2) is seen as a way of filling the gap left by the eventual depletion of fossil liquid and gaseous fuels, since, utilizing the electricity which might be more widely available from renewable, nuclear or other sources, it provides a route to convenient and easily integrated substitutes for the fossil fuels . An obvious likely impediment to the widespread use of hydrogen itself is lack of public acceptability in view of its well-known ready inflammability. It is worth recalling, however, that town gas was a roughly equal mixture of hydrogen and carbon monoxide and for many decades provided an acceptable energy store and fuel.

Energy distribution

It was suggested above that the inherent characteristics of most renewable sources of energy, particularly their widespread availability, low intensity and small scale, makes them ideally suited to a highly dispersed type of deployment whereby power is supplied to the immediate locality. Recent trends in power generation, especially in MDCs, have tended to move in the opposite direction, towards economies of scale and reliance upon efficient distribution networks for supplying consumers (table 5.1). Recent policy in the UK has been to build coal-fired stations (of 2 GW capacity) close to ample supplies of fuel, and nuclear stations (of 1.4 GW capacity) sited away from large centres of population. A comprehensive national grid carries electrical power to the remotest parts of the country (fig. 5.3). No doubt any significant development of electricity production from renewable sources, which would in most cases be variable and intermittent, would be fed into the grid and used to save fuel consumption in ordinary power stations.

One of the attractive aspects of electrical power is the comparative ease with which it can be

transferred in continuous fashion with only moderate losses over many hundreds of kilometres. Since transmission lines necessarily have a finite electrical resistance (unless at very low temperatures where they may become superconducting), some power, ΔP, is always lost by ohmic heating according to the relation

$$\Delta P = i^2 rl$$

where i is the current flowing and r the resistance per unit length l of the transmission cable. If a quantity of power, P, is being transmitted at a voltage, V, then the current flowing in the cable is

$$i = P / V$$

hence the power loss, ΔP, is proportional to the resistance of the cable, rl, and inversely proportional to the square of the transmission voltage, V. Since there are both economic and engineering constraints to the maximum thickness (giving minimum resistance) of cable that can be used, transmission voltages are made as high as insulation limits will allow. Power is generated at a few kV, transformed up as high as 400 kV for long-distance transmission, then transformed down again in successive steps for local distribution and ultimately consumption at 240 V (fig. 5.3). Alternating current is used for ease of voltage conversion. In the UK, about 8% of electricity generated is lost in transmission and distribution.

Crude oil and natural gas are transmitted over great distances by continuous pipeline; coal, crude oil and some liquefied petroleum gas (LPG) are similarly transported in bulk loads by ship. If it were to come into widespread use, hydrogen is capable of similar transportation. Apart from electricity, renewable energy in the form of heat may be transferred only over relatively short distances, because of unavoidable losses. It is generally uneconomic to transport fuelwood and

Table 5.1 Electricity generation by fossil fuel fired power stations in the UK

Year	Approximate number of power stations	Total generating capacity (GW)	Average efficiency (%)
1944	340	12	21
1953	310	19	23
1962	260	36	28
1971	200	55	30
1980	160	59	32
1989	80	60	34

Source: DEn Digest of UK Energy Statistics (annual)

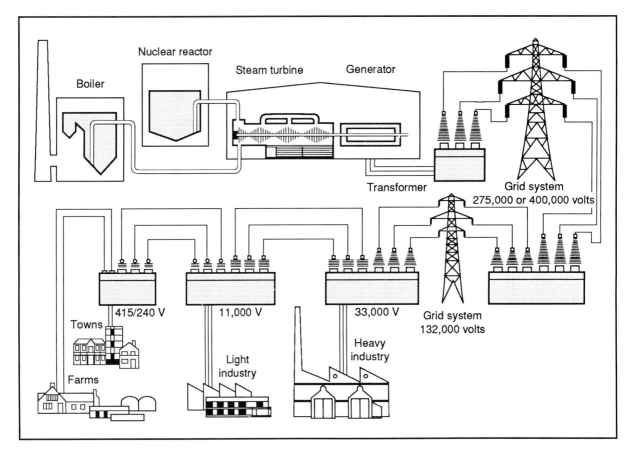

Figure 5.3   The network of the national grid provides the link between generators and consumers of electricity.
Source: National Grid

other forms of biomass over large distances unless converted to fuels, such as alcohol, of higher energy density.

Points to consider

- How valid and how useful is the distinction sometimes made between renewable sources of energy as income and non-renewable sources as capital?

- Are renewable energy sources desirable in principle?

- What are the main applications of energy storage?

- In which particular applications could energy storage (a) help to conserve fossil fuels (b) reduce environmental pollution?

- Does electrically driven road and rail transport (a) create environmental pollution (b) make more efficient use of fossil fuels?

- Suggest what further developments are needed to make a hydrogen economy a practical possibility.

- What impact would a hydrogen economy have on your daily routine?

- How necessary to our present way of life is the national grid?

- Compare the advantages and disadvantages of local as opposed to central generation of power.

- To what extent do systems of energy distribution result in additional use of (non-renewable) resources and pollution of the environment?

6 Energy from the Sun

• In what different ways can we make more use of the enormous flux of solar energy?

• Can solar energy be useful in cloudy regions or at high latitudes?

• Is solar energy particularly appropriate to LDCs?

• Is it useful mainly as a source of heat?

• What environmental impacts are involved in its exploitation?

As can be seen from fig. 5.1, the amount of solar energy reaching the Earth's surface exceeds current world consumption of energy by more than four orders of magnitude. The difficulty lies in harnessing this flux, whose peak value at sea level is 1 kW m^{-2}, with a global average of about 200 W m^{-2}. At any particular location the amount of solar radiation received will depend, predictably, on the time of day and season of the year as well as geographical latitude (fig. 6.1). The weather provides an unpredictable element; under clear skies most of the radiant energy will be direct with no more than about 15% diffuse, but the latter component will obviously rise to 100% if skies are overcast. The low flux density implies that, unless energy storage is involved, a large area of collection is required to deliver significant amounts of power. Efficiency can be optimized by orientating the collecting surface perpendicular to the solar beam.

Exploitation of solar energy must therefore contend with two principal drawbacks: its relatively low power density and its variable availability. For the purposes of this discussion, the possibilities for the conversion of radiant energy will be divided into three: photothermal, concerned with the active and passive collection of solar radiation as heat; photoelectric, its direct conversion into electricity; and photosynthetic, whereby solar energy is stored through biochemical reactions in green plants.

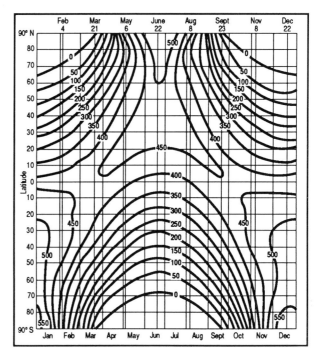

Figure 6.1 **The variation with latitude and time of year of the maximum amount of solar radiation (in kW m^{-2}) received on a horizontal surface at sea level. The lack of symmetry between the two hemispheres is due to a slight ellipticity of the Earth's orbit, which results in the shortest distance to the Sun (perihelion) occuring on 4 January and the greatest distance (aphelion) on 5 July.**
Source: Briggs and Smithson (1985)

Photothermal applications

Building design

You will surely have encountered the stifling heat that quickly builds up in an unventilated attic or loft room, conservatory or greenhouse on a bright summer's day and you may once have experimented with burning holes in a sheet of paper with a small magnifying glass! These two types of experience illustrate respectively the passive and active approaches to the utilization of the Sun's heat.

The art of making the indoor environment of buildings as comfortable as possible through all seasons of the year by means of passive solar design has a very long history, stretching back at least to Ancient Greece. Present-day approaches aim to modify the arrangement and composition of all the components of a building – walls, floors, roof, windows – to make the most effective use of the incident solar radiation falling upon it so as to minimize the need for supplementary space heating during the winter half of the year, whilst also taking precautions to avoid overheating during the summer months.

In a *direct solar gain* building, a large window area is located in walls facing within about 30° of south (in the northern hemisphere) with any north-facing window area reduced to the absolute minimum. The incoming solar radiation is absorbed and stored within the fabric of the internal environment, which must be well insulated from losses to the outside world. For example, the large south-facing windows will need moveable insulation to restrict night-time and winter season heat losses, whilst evergreen vegetation planted on the northern side of the building could provide an effective shield against cold winds. Conversely, deciduous trees on the southern side would allow the free passage of valuable winter sunshine whilst screening out the perhaps too intense radiation of summer. Projecting eaves or window shades may also be used to provide shields for summer radiation (fig. 6.2).

A direct solar gain building relies upon natural circulation of the absorbed radiation; taking into account the incidental gains from the waste heat output of lighting, electrical appliances and the occupants, it will, if sufficiently well insulated, have a minimal (or even zero) requirement for supplemental space heating. A number of houses of passive solar design have been constructed at Pennylands and at Great Linford in Milton Keynes with the aim of evaluating different distributions of glazing and levels of insulation (figs 6.3 and 6.4). While adding less than 1% to construction costs, they have achieved reductions in space heating requirements of up to a half when compared with contemporary dwellings built to more conventional designs. In countries where the winter climate is more severe than in the UK, it is advantageous to incorporate some form of inter-seasonal heat storage or to employ panels of solar collectors.

The most straightforward method of *indirect solar gain* involves the addition of a conservatory onto an unshaded wall with a southern aspect. This was a feature popularized by the Victorians and is now seeing something of a revival. As well as providing a useful and pleasant extension of the living space, it can make a significant contribution (up to 10 GJ) to annual space heating needs. For multi-storey office blocks, a solar atrium may fulfil a similar function, as, for example, at the Department of Energy's London headquarters.

The *Trombe wall* is not dissimilar in principle to the conservatory and can be thought of as a solar operated 'night' storage heater (fig. 6.5). A thermally massive blackened wall is placed behind glazing on the south-facing side of the dwelling.

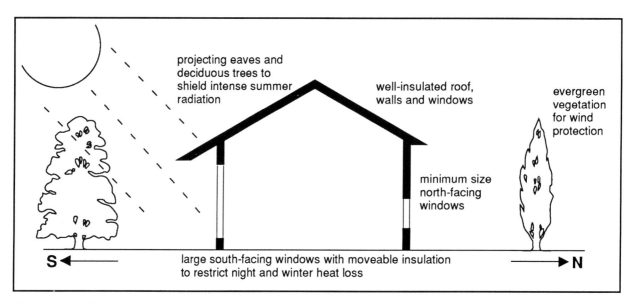

projecting eaves and deciduous trees to shield intense summer radiation

well-insulated roof, walls and windows

evergreen vegetation for wind protection

minimum size north-facing windows

S ← large south-facing windows with moveable insulation to restrict night and winter heat loss → N

Figure 6.2 Passive solar design for a house.

Figure 6.3 Houses at Pennylands, Milton Keynes, are placed with small windows in north-facing and large windows in south-facing walls. These houses incorporate other energy-conserving features.
Photo: ETSU

Figure 6.4 South elevation of the Linford houses.
Photo: ETSU

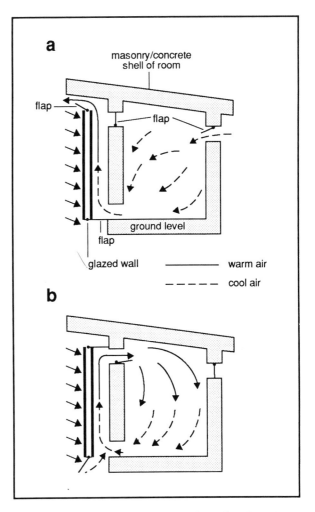

Figure 6.5 A schematic illustration of a Trombe wall showing the mode of operation for a) cooling and b) warming.
Source: Open University (1984)

As well as radiant heat delivered to the living space, warmed air circulates by natural or forced convection through a system of ducts. Provision may be made on the one hand to minimize heat losses at night and on the other to allow for cooling on hot days. Alternatively, the Trombe wall can be filled with water to achieve the required high thermal capacity and the entire structure may be relocated as a roof pond, provided that the load can be borne by the building.

Thermal collectors
In addition to space heating requirements, the occupants of both domestic and industrial buildings generally have a significant demand for hot water; active solar energy collection systems can contribute to both of these needs.

The simplest type of *flat-plate collector* is conveniently mounted on a south-facing roof (fig. 6.6). Its essential features include:

i) a surface which is a good absorber of solar radiation (and preferably also a poor emitter of radiation);
ii) a series of pipes in good thermal contact with the absorber plate through which a circulating fluid (usually water) may transfer the heat directly to the point of use or to a store;
iii) a transparent cover (glass or plastic) to reduce heat losses from the upper surface of the collector;
iv) adequate thermal insulation on any other exposed warm surfaces.

The construction of the panel must be sufficiently robust to withstand wind, rain and particularly wide fluctuations in temperature (some early installations of this type suffered rapid deterioration, and contributed to a reaction against this type of system). The possibilities of internal corrosion and the freezing of the working fluid must also be guarded against. For optimum efficiency the angle of tilt of the panel should be seasonally adjusted but the natural pitch of the

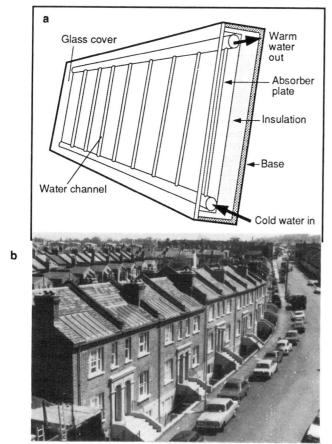

Figure 6.6 a) A solar flat-plate collector (solar panel)
b) Solar panels installed on the roofs of a rehabilitated Victorian terrace in Whately Road, East Dulwich.
Photo: ETSU

roof usually provides a satisfactory compromise. Although natural convection occurs, circulation is usually by means of a pump through a heat exchanger to a hot-water cylinder. For a typical house in the UK, hot-water requirements could be met by about 5 m² of such panels. Water temperatures of 40 to 50°C are readily achieved by this sort of design, but above 80 to 90°C heat losses exceed gains even for the most advanced types.

In the UK the widespread adoption of solar water heating systems is inhibited by a general lack of sunshine and a degree of mismatch between the demand for, and the availability of, hot water; such a system is unlikely to provide more than about 40% of annual requirements. Furthermore, because of the limited market, expected economies of scale resulting from mass-production have not been realized. The heating of outdoor swimming pools provides perhaps the best marriage of demand with availability. The use of water heating systems is, not surprisingly, much more

common in Mediterranean countries: compared with an incidence of about 1 in a 1000 households in the UK, France and Italy have 1 in 300 and Greece 1 in 60. In the exceptionally sunny climate of Cyprus, some 90% of households make use of solar water heaters, while in similarly favoured Israel, where all new buildings must by law be fitted with such a system, the figure is 2 in 3. Other countries where the technology is becoming commonplace include Australia, Japan and the southern USA.

The solar pond combines collection and storage of low-grade heat and is best suited to latitudes where the elevation of the Sun remains reasonably high throughout the year. It is a relatively simple construction (fig. 6.7) consisting of a shallow body of water about 2 to 3 m deep. The lowest layer contains dissolved salts in sufficient concentration that, even when it becomes hot, it is dense enough to remain confined to the bottom of the pond. Successively higher layers of water have lower concentrations of salts so that thermal convection is suppressed, the whole remaining stably stratified at all temperatures. The bottom of the pond is blackened to enhance absorption of solar energy and it may also be insulated to reduce heat losses by conduction; in this way temperatures of up to 100°C can be attained. The heat store can be tapped by drawing off the hot brine or by means of heat-exchange coils within the lowest brine layer. It may be utilized directly as process heat or, by means of a heat engine with a volatile working fluid, to generate electricity. Although of low efficiency, simplicity of design makes for a low-cost technology. An early installation of 7000 m² surface area near the Dead Sea in Israel generated 150 kW of electricity and a larger 5 MW unit has since been constructed.

Many developing countries have a high demand for fuelwood for cooking. The satisfaction of this demand is placing increasing stresses upon the local environment. Although the best approach in practical terms may be to try to improve the efficiency of conventional ovens (Foley et al., 1984), it is possible to design solar ovens which eliminate the need for fuelwood entirely. While a simple hot-box, having a window, blackened interior surfaces and good insulation, can be sufficient, efficiency can be greatly improved (and cooking times reduced) by incorporating reflecting mirrors, focussing Fresnel lenses or solar tracking into the design. However, more complex technology, especially when it results in greater

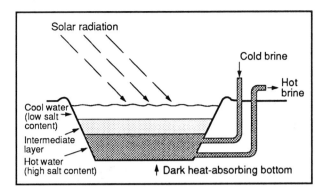

Figure 6.7 Schematic diagram of a solar pond.

initial costs or difficulties in operation or of maintenance, is likely to inhibit the take-up of solar cookers and to intensify any resistance to the abandonment of traditional practices. It is well to remember that important social and cultural dimensions are likely to attach to the preparation and consumption of food to which solar cookers may be poorly adapted, especially if the principal meal is normally prepared indoors or at sunset.

Further possible applications of solar energy absorption at relatively low temperatures include solar stills for providing fresh water and absorption-type refrigeration units for cool storage or air conditioning.

Somewhat higher temperatures of up to 200 to 300°C can be attained by using vacuum tube collectors, which eliminate heat losses through conduction and convection, with or without a system of mirrors (or Fresnel lenses) to focus a greater proportion of solar radiation onto the absorbing surface. A further gain in efficiency, at the cost of increased complexity of design, is possible if the focussing system follows the movement of the sun in either azimuth or elevation.

For still higher temperatures it is necessary to concentrate the solar energy collected over a considerable area at a single point. Parabolic dish collectors can reach temperatures of up to 900°C while 1400°C is achievable with a sufficiently large number of individual solar-tracking mirrors (heliostats) delivering solar energy to a central receiver. An experimental station of this type at Odeillo in southern France constitutes a solar furnace, having attained temperatures of 3000°C and an output of 1 MW of energy. In this type of device quite precise optical surfaces and alignments are called for and the heliostats must track the Sun in both azimuth and elevation. With

such systems it is possible to generate steam having a temperature comparable with that used in conventional power stations, and so electricity generation with a similar efficiency is feasible.

In the Mojave Desert near Barstow in California five 30 MW stations, or *solar power towers*, have been built and more are planned; each has some 90,000 m² of heliostats occupying about 40 ha. Their costs per MW are similar to those of nuclear stations and their power output can usefully offset peak demands from air-conditioning units. Constraints on the optical quality, rigidity and mobility of the most distant mirrors suggest that 30 MW is the practical upper limit on the size of such thermal-electric installations. It is also apparent that a total output on the scale of a conventional power station would require an area of a few km² to be devoted to the collection of sunlight.

Photoelectric conversion

It was shown in the preceding section how solar energy could be collected and focussed or stored to an extent where it could be used, by driving a 'heat engine', to generate electricity at a low but locally acceptable efficiency. It is also possible, and, remembering the limitations of the second law of thermodynamics, at first sight preferable, to convert sunlight directly into electricity through the interactions of solar photons. Unfortunately, the simultaneous maximization of the output of both current and voltage provides a mutually incompatible design requirement which, together with the inevitable less-than-ideal nature of practical materials, limits conversion efficiencies to around 30% in the laboratory and 15–20% in practice.

Photovoltaic cells based on silicon, doped with impurities such as phosphorus and boron to give n-p type semiconductor (see glossary) properties, were first developed in the 1950s for use in satellites as a clean, lightweight, safe and reliable power source; their inordinate expense (about $200,000 per peak kW) was not in that context a significant drawback. To date, advances in both materials, design and manufacturing techniques have reduced costs by nearly three orders of magnitude and further cost reductions are to be expected. As a result, applications to devices requiring a very small power input, such as calculators and watches, are now commonplace.

These *solar cells*, as they are popularly known, work equally satisfactorily in bright or diffuse sunlight as well as artificial illumination.

Individual cells, being based on single crystals of silicon, are limited to a maximum diameter of about 10 cm. Up to 100 cells are combined, using appropriate combinations of series and parallel connections, into *modules* about 1 m² and having a convenient power output (about 50 W). They are robust and easily handled. Their power output is designated in terms of 'peak watts', which assumes an incident irradiance of 1000 W m⁻², and would be about 15 mW cm⁻² at 25°C, but declining with increasing temperatures. Modules in turn are linked together into larger *arrays* to deliver any desired combinations of current and voltage (fig. 6.8). In order to deliver 1 kW on average in the UK, arrays rated at five or six peak kilowatts would be needed, but the figure could be as low as three in sunnier climates.

The costs of solar arrays have already fallen to the point where serious attention is being given to limiting the expense of necessary ancillary equipment such as rigid and durable supports, banks of rechargeable batteries for power storage against periods of non-generation and, where necessary, inverters to convert the DC solar cell output to conventional AC. Very large arrays can be constructed to deliver power outputs in the tens of MW range but the requirements for land are comparable to those of the solar power towers described above. Most current applications are at a smaller scale and are already economic: applications to calculators and watches were mentioned above; others include remote data monitoring, navigation warning and telecommunications relay stations, and maintaining the charge of batteries for emergency equipment. In the slightly higher power range of 0.5 to 150 kW, which would include remote small communities and radar installations, for example, storage batteries are usually required, but solar cells may still be more economic than the conventional alternative, a diesel generator.

The constant availability in space of solar energy of higher quality than at the Earth's surface has prompted some interest in the practical possibility of very large arrays of solar cells mounted in geostationary orbit above a ground-based receiving station. In one such scheme, the collecting satellite would support some 10¹⁰ solar cells on a light framework about 50 km² in surface

Figure 6.8   **Arrays of solar cells on display at Machynlleth, Powys.**
Photo: Centre for Alternative Technology

area, and a 1 km diameter antenna could transmit some 5 GW of microwave power down to a 10 km diameter receiving antenna. Apart from the obvious technical difficulties and present high cost of operations in space, the possibilities for military applications of intense beams of microwave energy would need to be borne in mind.

Photosynthesis and biomass

The amount of stored energy in the form of biomass resulting from photosynthetic processes exceeds the world's annual consumption of energy by an order of magnitude. It is an important energy resource not only because of its large size but also because it constitutes a natural store and is readily capable of conversion into a variety of gaseous, liquid and solid fuels – important attributes which are deficient in most sources of renewable energy. Moreover, like natural hydrocarbons, they may be used as a feedstock in the derivation of a wide range of organic chemical products.

Plants containing chlorophyll are able to utilize (solar) photons to split molecules of water (thus liberating oxygen to the environment) and then combine the hydrogen with carbon dioxide to build up a variety of carbohydrate molecules. The net effect of the complex series of reactions involved may be summarized as

$$6\,CO_2 + 6\,H_2O + \text{solar energy} \rightarrow C_6H_{12}O_6 + 6\,O_2$$

Some of the stored chemical energy will be needed by the plants themselves, but most is available to supply food, or for conversion to other forms of energy, for the whole of the biosphere. Micro-organisms are responsible for the breakdown and recycling of all of this organic material within a relatively short time after its production. Photosynthesis can take place under conditions of diffuse as well as direct sunlight but is limited to low average efficiencies in the range 0.5 to 1.5%, although 2.5% or more can be achieved by some tropical plants.

The collection of wood from natural vegetation for use as a fuel has a very long history and remains the most important energy resource for many of the LDCs, and their rural populations in particular. While wood is, in principle, simple and economical to gather, store and use, it is also bulky in relation to its energy content and its combustion gives rise to atmospheric pollution. Moreover, its collection in increasing volumes, coupled with other significant demands (including commercial logging and the clearance of forest for agricultural development) has led to serious ecological degradation. Because the collection of fuelwood is traditionally a task for women and children, the increasing time that has had to be devoted to this work has, through the consequential neglect or simplification of other tasks, had adverse effects on the physical and social well-being of rural families.

The mitigation of these negative aspects can be approached through the use of the various possible conversion technologies as well as the employment of alternative sources for the supply of biomass material.

Biomass conversion

As mentioned above, a unique advantage of biomass resources of all types is their suitability for conversion into a variety of gaseous, liquid and solid fuels (fig. 6.9 and table 6.1). Different pathways may be selected depending on the nature of the source material and the quantities available, as well as local environmental factors and demand for the end-product. The various technologies can be conveniently separated into *dry* and *wet* processes.

The most straightforward way of using biomass is by direct *combustion*. The large relative bulk in relation to energy content generally makes its transport over long distances in MDCs un-economic, while in LDCs this factor, together with an increasing scarcity of supply, adds significantly to the burdens of daily existence. Air-dried wood has an energy content of some 15 MJ kg^{-1}, about half that of bituminous coal; kiln-drying can increase this value by about one-third, at the cost of an input of energy.

Pyrolysis involves the heating of plant material in the absence of oxygen to give a variey of gaseous, liquid and solid fuels. The conversion of wood into charcoal is simple and widely practised in

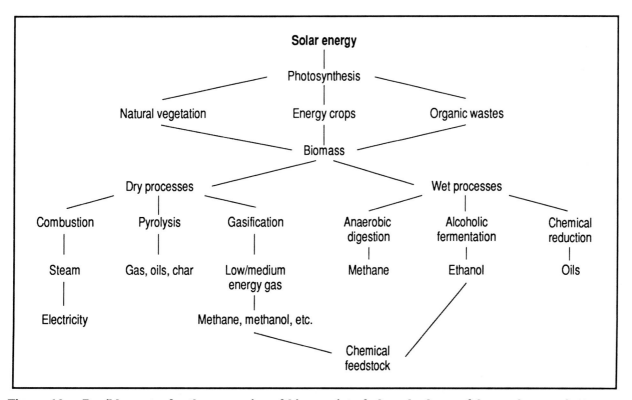

Figure 6.9   Possible routes for the conversion of biomass into fuels and other useful secondary products.

Table 6.1 Approximate energy content of biofuels expressed as gross calorific value in MJ kg⁻¹

Wood: green	5–10
seasoned	12–15
oven-dried	20
Charcoal	28
Char	20
Peat (dried)	10–15
Straw	16
Bagasse	6–9
Cattle dung (dry)	~12
RDF/WDF (refuse/waste derived fuel) 15 – 20	
Ethanol	28
Methanol	21
Biogas	20 (20 MJ m⁻³)
SNG (Substitute natural gas)	25–35
Hydrogen	120 (12 MJ m⁻³)

Source: adapted from Slesser (1988); Twidell & Weir (1986)

LDCs. It has the economic advantage of reducing the bulk density of the fuelwood, so making transport easier and cheaper, while at the same time increasing the calorific value by a factor of about two, which is then comparable with bituminous coal and therefore suitable for high-temperature uses. However, the environmental costs are high: much of the original energy content of the wood is lost in the conversion process, and the volume reduction may be as much as ten times. A technologically more complex approach can yield a mixture of gaseous, liquid and solid hydrocarbons, whose proportions and composition vary with temperature (fig. 6.10).

Gasification techniques are intended to maximize gas production from pyrolysis at high temperatures. Small amounts of air blown into the reaction vessel give *fuel gas*, a mixture of nitrogen, carbon monoxide and hydrogen, of rather low calorific value (about 5 MJ m⁻³). Using oxygen instead of air eliminates the inert nitrogen fraction so increasing the calorific value by a factor of two to three. A further possibility involves the use of hydrogen (chapter 5), which may be conveniently generated within the reaction vessel by the action of carbon monoxide on steam. The resulting mixture of methane and carbon dioxide, or *substitute natural gas* (SNG), has a calorific value of up to 35 MJ m⁻³.

Human and animal excreta are ideally suited to the process of *anaerobic digestion*; green plant matter may also be successfully used but drier potential feedstocks, such as refuse, straw and sawdust, digest very slowly. The high content of putrescible material in present-day household refuse means that methane generation can present a long-term hazard near to waste tips; alternatively, the tip may be designed so that the methane provides a valuable fuel for a nearby consumer (fig. 6.11). The natural bacterial decomposition processes are much more rapid in an aqueous medium maintained at an optimum temperature of about 35°C. In temperate climates this requires some input of heat for much of the year, but a small number of commercial installations are operating on livestock farms in the UK, for example. The residue from the digestion process is largely devoid of both odour and harmful bacteria, and, because of its high nitrogen content, is a valuable fertilizer and soil conditioner, or may be used as a protein supplement in animal feed.

At a smaller scale, the technology has been extensively deployed in LDCs where ambient temperatures are sufficiently high that no supplementary heating is needed (fig. 6.12). In China, disposal of domestic sewage was an important factor behind the construction of some half-million digesters, although to supply adequate quantities of methane for both cooking and

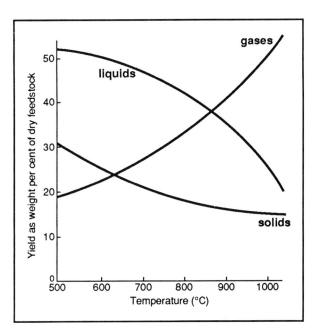

Figure 6.10 The effect of temperature on the yield of products from the pyrolysis of biomass.
Source: WCE (1979)

Figure 6.11 Shell-type boiler fired by landfill gas.
Photo: ETSU

lighting in the local community, the contribution of a number of livestock is necessary. In India and elsewhere, the digestion of animal dung to provide both fuel and fertilizer has proved a much more efficient option than the practice – especially prevalent where fuelwood is in short supply – of using dried dung directly as a fuel. Methane is also a cheaper alternative to paraffin (kerosene), which was often a practical option for cooking and lighting before the oil crisis of 1973. However, even such a relatively simple device as a methane digester requires a minimum amount of expertise in both construction and maintenance as well as a degree of co-operation amongst members of the community in day-to-day operation; it may also meet resistance to the extent that it supplants traditional practices.

The *fermentation* of sugars by yeasts is a very long established procedure in the production of alcoholic beverages, but its development on a large scale to provide ethanol as a liquid fuel is more recent (Kovarik, 1982). The most suitable starting products are plants with a naturally high content of sugars, such as cane, beet or fruit; the starch and cellulosic content of corn and potatoes is also suitable as feedstock after an initial enzymatic or chemical hydrolysis into sugars. In little more than a day of fermentation, the alcohol content of the 'brew' will be between 6 and 10%; unfortunately, the following distillation stage can have a high energy demand, possibly in excess of the energy content of the resultant ethanol fuel (22 MJ l^{-1}). Car engines can be easily modified to run on ethanol or, without modification, on a 1:9 ethanol–petrol blend known as 'gasohol'. A large programme of ethanol production, mainly from sugar cane and its residues, was embarked upon by Brazil in 1971 with the aim of reducing dependence upon imported crude oil and the consequent pressure on limited reserves of foreign currency. Methanol, with a slightly lower calorific value, is similarly suitable as a fuel for transport; it may be derived from cellulosic feedstocks (usually by gasification techniques) from, for example, logging and sawmill wastes.

Chemical reduction in alkaline solution of plant and animal wastes is a feasible but less developed approach. Depending on the chosen levels of pressure and temperature, the yield can be either gas of high calorific value or a mixture of oils.

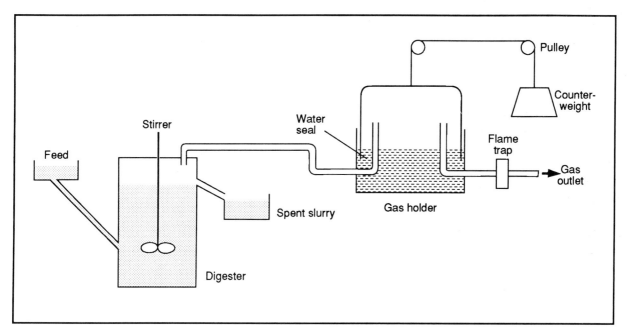

Figure 6.12 Schematic diagram of an anaerobic digester for the production of methane gas.
Source: Dunn (1978)

Sources of biomass

Biomass for energy production may be obtained from three types of source – natural vegetation, plants grown explicitly for the purpose (often referred to as 'energy crops') and organic waste products of many different origins (fig 6.9).

Natural vegetation, and wood in particular, is the oldest source of energy to be exploited by the human race and it still constitutes the single most important one for many LDCs today (fig. 6.13). Indeed, it accounts for as much as 60% of energy consumption in Africa and about 6% in the world. It has the obvious advantages that it is usually free (though, as mentioned above, its collection may well involve costs other than of a monetary kind) and widely available, even, or perhaps especially, to remote communities. In many parts of the world, however, the exploitation of natural sources of firewood far exceeds their capacity for regeneration, with consequent serious ecological and environmental disruption and deterioration. Some observers consider that this depletion and extinction of fuelwood sources in less developed regions constitutes a more pressing energy crisis than the dwindling fossil fuel reserves of the more developed world (see, for example, Eckholm *et al.,* 1984; Munslow, 1988; Soussan, 1988).

It is of course possible to reclaim and replant areas denuded of vegetation but, where the intention is principally to provide fuelwood, at least twice as great a return can be obtained if species are selected for their rapid and regenerative growth. In temperate regions, for example, trees such as alder, hazel, poplar, sycamore and willow grow at a rate of 2 to 3 m a^{-1} and respond well to *coppicing*, whereby the regenerated shoots are cut back to a stump close to ground level on a three to five year rotation. Because of the well-established root system, the regenerated growth is much faster than for a standard tree (fig. 6.14). In this way, the equivalent in energy terms of some 150 GJ ha^{-1} a^{-1} can be harvested. In warmer regions, various species of eucalyptus are favoured for the same purpose in energy plantations; yields are generally higher, up to 300 GJ ha^{-1} a^{-1}, while the extensive root systems are helpful in the control of soil erosion.

One of the most familiar of *energy crops* is sugar cane, which has a higher than average photosynthetic efficiency (about 2–3%) and yields by direct fermentation up to 70 l t^{-1} of ethanol; sweet sorghum is an alternative tropical crop having similar characteristics. Most staple food crops are rich in starch, but this must be hydrolysed into sugar before fermentation. Potatoes and maize (corn) are examples of suitable temperate crops, while cassava can be exploited on hotter, drier soils of poorer quality than are tolerated by sugar cane.

It must be borne in mind that energy crops and plantations may be in competition for land with conventional agriculture and those who practise it,

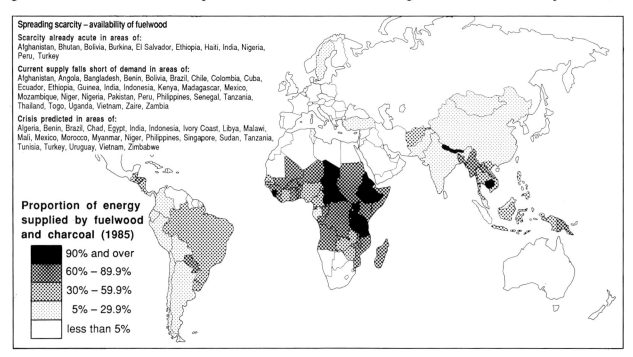

Figure 6.13 **The fuelwood crisis.**
Source: Seager (1990)

Figure 6.14 Two-year-old willow coppice at Long Ashton. Aberdeen University trial, August 1988.
Photo: ETSU

whether large landowners or subsistence peasants. Possible environmental, social and cultural impacts of energy production schemes must not be overlooked. Fast-growing water plants might avoid some of the difficulties; in tropical latitudes, for example, the water hyacinth (*Eichhornia crassipes*) is capable of doubling its volume in little more than a week.

The need for conversion technologies may be avoided with a small percentage of plants that directly produce hydrocarbons of high molecular weight. Examples include latex from *Euphorbia* species and oils from sunflowers, palms and jojoba. It should also be pointed out that these plant products, as well as the results of the various conversion technologies on other sources of biomass, constitute not only a store of energy but also a chemical feedstock, a possibly important future alternative to petroleum or coal.

In principle organic wastes possess two advantages over energy crops – the use of waste from other enterprises means that competition for resources is avoided and, since the wastes will in any case need to be disposed of, environmental impact is reduced, possibly with financial gain. The use of human and animal wastes, especially in LDCs, has been described above. In MDCs, domestic refuse invariably presents a disposal problem – the UK, for example, produces 26 Mt of non-hazardous domestic and commercial waste each year – but such waste has a sufficiently high calorific value to be of use as an energy source. In the UK landfilling of wastes is the preferred and still the most cost-effective disposal option, but incineration, often with energy recovery, is more widely practised elsewhere in Europe. If this latter route is to be adopted, then all metal and glass

material must be removed, either after collection or by arranging for segregated disposal by the householder. The energy content of the residual waste is most efficiently utilized by combustion in boilers feeding district heating networks rather than by supplying steam for the generation of electricity (chapter 9). The boilers must be designed to handle the bulky heterogeneous fuel and to limit the risks of corrosion and atmospheric pollution, particularly from SO_2, HCl, dioxins and furans.

In the UK, Nottingham has the largest of a very small number of city-based schemes relying principally on the incineration of municipal refuse. Each year, 175 kt of refuse are used to supply space heating and hot water to some 7000 homes, offices and shops and to generate 2.6 MW of electricity. The incinerator at Edmonton, North London, consumes more than 1 kt of locally collected municipal refuse per day but generates only electricity, sending some 25 MW into the National Grid. A new facility is under construction in Wolverhampton which will generate 22 MW of electricity from the incineration of scrap tyres at a rate of 12 million a year, the permanent disposal of which would otherwise present many difficulties.

An alternative option, which allows for storage of the fuel and its use by a variety of industrial consumers, involves conversion into small pellets of refuse derived fuel or RDF (also known as waste derived fuel, WDF) having a lower calorific value than coal (about 60%) but satisfactory as a fuel in combination with coal, or alone in equipment designed to cope with its different physical and chemical characteristics. The UK has six such plants capable of supplying more than 130 kt a^{-1}, but producing much less because of limited demand. The output of one plant at Byker on Tyneside, for example, is used in a local district heating scheme.

As mentioned above, even when refuse has undergone conventional landfilling, some of its energy content can still be available in the form of landfill gas, a mixture of up to two parts of methane to one of carbon dioxide. The lengths of time over which useful yields of gas may be obtained are considerably increased if the tip is designed from the outset for gas recovery. A large landfill site in a worked-out brick clay pit at Stewartby, near Bedford, for example, supplies gas to fire nearby brick kilns as well as generating 0.8 MW of electricity.

Dry wastes from agriculture and forestry, such as straw, brash and wood chippings, have a useful energy content but, when compared with domestic refuse, the economics are unfavourably affected by the higher costs of collection and transport, unless specially compacted into high-density bales. Rural enterprises between the Thames and the Humber are the most likely beneficiaries from the energy content of the 7 Mt of surplus straw produced each year in the UK. For example, heating the buildings of Woburn Abbey with an 800 kW boiler consumes more than 400 t a^{-1} of chopped straw.

The energy content of wet agricultural wastes is most conveniently realized through processes of anaerobic digestion. In the UK there is an annual surplus of some 70 to 80 Mt, principally from dairy and beef cattle, with smaller contributions from pigs and poultry, which could yield between 2 and 3 km^3 of biogas. As mentioned above, there are at present both economic and technical disincentives affecting the exploitation of this resource. However, in a new operation at Eye, Suffolk, it is proposed to generate 14 MW of electricity from the incineration of 130 kt a^{-1} of poultry litter.

Points to consider

- How serious a drawback is the variability of solar energy?

- To what extent can existing dwellings benefit from passive solar gain?

- Assess the relative benefits and costs of the large-scale deployment of flat-plate collectors in the domestic sector.

- How relevant are attempts to achieve very high temperatures by bringing sunlight collected over a large area to a point focus?

- Would an energy supply based on solar (photoelectric) cells be entirely non-polluting and totally without environmental impact?

- What do you consider to be the most appropriate applications for solar cells?

- What are the principal advantages of biomass compared with other renewable energy sources?

- To what extent does the use of biomass as a source of energy have either positive or negative environmental impacts?

- What factors in different countries determine the most readily available sources of biomass and the most appropriate conversion technologies?

- How do different conversion technologies and end-uses vary in the efficiency with which they utilize biomass energy?

- Consider the implications for LDCs of using alternative sources of energy to either charcoal or fuelwood.

- What is the approximate energy content of the waste materials of agricultural and domestic origin produced in the UK each year? How does this compare with the UK's annual energy demand? Can you suggest why these sources of energy are under-exploited?

- Does the use of solar energy automatically protect the environment from degradation and despoliation?

7 Energy from the sea

- How much energy does the sea contain?

- What forms is it in and how can it be harnessed?

- What environmental impacts might result from using it?

SO MUCH FOR WAVE ENERGY! I DON'T SEE ANY!

The world's oceans, covering some 70% of its surface, contain vast stores of energy but, at the same time, they are a remote, alien and often hostile environment. Their energy resources – thermal, kinetic and gravitational – are neither in a readily usable form nor near to sources of demand. Technical approaches to harnessing this energy have therefore understandably concentrated on its conversion to electricity at coastal or near-shore locations.

There are three principal energy stores which have begun to be exploited. Ocean thermal energy conversion (OTEC) utilizes the fact that solar energy is absorbed by the surface waters of the oceans, which in low latitudes are permanently warm. The thermal energy store can be tapped by making use of the temperature difference between these warm surface and deeper cold waters. A second energy store derives indirectly from the fraction of solar energy that drives the atmospheric circulation; winds blowing over the sea transfer in turn some of their energy to the waves they generate. Finally, the level of even a calm sea varies regularly and predictably under the combined gravitational attraction of the Moon and the Sun. This tidal energy store can also be tapped at certain favourable locations.

The oceans also have other potential sources of energy, including the kinetic energy contained in

tidal and wind-driven currents and osmotic energy resulting from a salinity gradient such as when fresh estuarine water meets the saline waters of the open sea. However, there are no serious proposals to tap such sources at present. It might be feasible to harvest biomass, such as giant kelp, from the near shore and subject it to the types of process discussed in chapter 6 above. It may also become feasible to tap submarine geothermal sources in ways similar to those described in chapter 8 below.

While ocean thermal energy is permanently available, energy from waves and tides is variable in both magnitude and frequency, predictably so for tides.

Ocean thermal energy conversion (OTEC)

A vertical profile of mean annual temperatures in the Atlantic Ocean (fig. 7.1) demonstrates the presence, within about 30° of latitude from the equator, of a well-developed temperature contrast between warm surface waters and cold deep waters. This is the *permanent thermocline*. Where this temperature difference (ΔT) is greater than about 20°C (fig. 7.2), the warm surface waters can be used to evaporate a working fluid of suitably low boiling point, such as ammonia or CFC, in a boiler. The vapour is used to drive a turbine and then condensed in a heat exchanger by the deeper

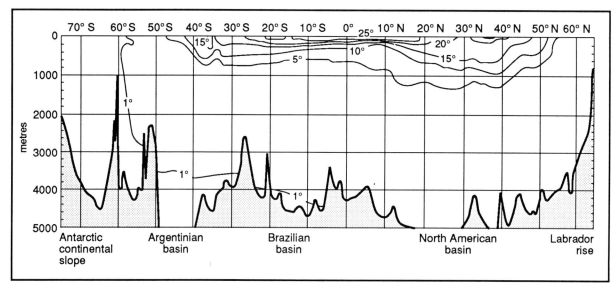

Figure 7.1 The variation of temperature with depth in the Atlantic Ocean. Isotherms in °C.
Source: Sverdrup *et al.* (1942)

cold water before being returned to the boiler (fig. 7.3). Alternatively, it is possible, though technically a little more difficult, to use seawater itself as the working fluid by evaporating it under a sufficiently reduced pressure. For generation of electricity, the turbine must be quite large because of the relatively low pressures involved, and the condensers larger still, but the overall efficiency is not much above 20%. A significant fraction of the power generated must be used to pump the necessarily large volumes of cooling water up from depths of more than 500 m.

Although many engineering difficulties are eased for installations based at the shore or in shallow coastal waters, cold deep water might then have to be pumped from tens or even hundreds of kilometres out to sea beyond the edge of the continental shelf. This particular problem is

avoided in an alternative approach which envisages a free-floating construction on the high seas (fig. 7.3). Here the capital costs would be even greater and there could be difficulties in transmitting the electrical power to land or in using it on board to extract economically valuable elements from seawater, for example, or to electrolyse seawater to produce hydrogen fuel. On the other hand, the fuel is free and permanently available, and environmental impacts, relating mainly to the effects of bringing cold oxygenated nutrient-rich water to the surface, could be beneficial to an enterprise such as fish farming.

The most favourable locations for OTEC plants appear to be remote (volcanic) islands in tropical oceans with steeply shelving coastlines; in the shallow seas close to the UK the thermocline is both too weak and seasonally variable. A 50 kW

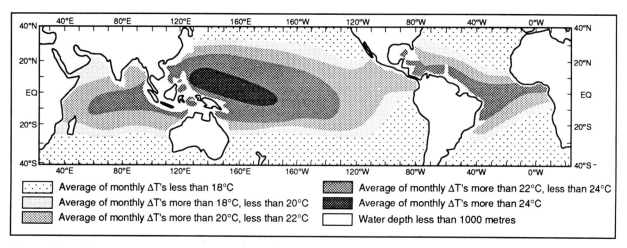

Figure 7.2 The average temperature of the surface waters of the oceans.
Source: IEA (1987)

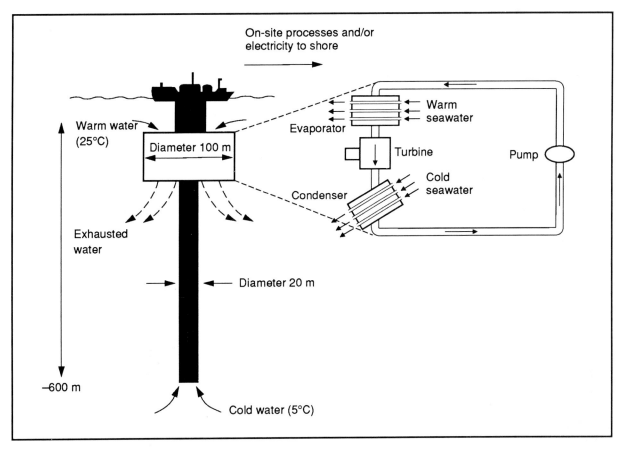

Warm water (25°C)

Diameter 100 m

Exhausted water

Diameter 20 m

−600 m

Cold water (5°C)

On-site processes and/or electricity to shore

Evaporator

Warm seawater

Turbine

Pump

Condenser

Cold seawater

Figure 7.3 A possible design for an OTEC plant for the generation of electricity. A plant generating 400 MW of electrical power would require a platform weighing 200 kt pumping 10 Mt of cold seawater per hour. Source: Taylor (1983)

(rated) pilot plant based on ammonia was successfully operated on Hawaii in 1979 although more than 80% of the power generated was used for pumping; plans to construct a 40 MW full-scale version were subsequently abandoned in the light of downward trends in oil prices and the advent of the Reagan administration. A pilot 100 kW installation of Japanese design and construction began operation in 1981 on the shore of the island of Nauru in the equatorial western Pacific (Marjoram, 1982). Using CFC-22 as the working fluid, 1400 t h^{-1} each of warm and cold water were used, the cold inlet pipe being 1 km in length. The amount of surplus power fed into the local electricity grid was 10 kW. All of the currently proposed or constructed plants are relatively small and shore-based; the construction and operation of much larger free-floating OTEC platforms is still some way into the future.

Power from the waves

The amount of energy transferred by the wind to the sea surface in the form of travelling waves depends on the strength of the wind, its fetch (or path length over the sea), and the length of time for which it blows. The energy possessed by a wave is partly kinetic, due to its forward motion, and partly potential, due to the wave height. The power per unit length in W m^{-1} of an idealized small-amplitude sinusoidal wave varies directly with the square of the wave height, H, as given by the equation

$$P = \rho\, g^2\, T\, H^2 / 32\, \pi$$

where ρ is the density of seawater and T the period of oscillation of the wave. For a typical sea state involving waves of varying heights and periods this approximates to P = 550 $T_e H_s^2$ W m^{-1}, where T_e is the period of the most powerful waves and H_s is the significant wave height. The energy in any group of waves travels at half the speed, c, of the individual waves which in turn depends on the wavelength λ as

$$c^2 = g\, \lambda / 2\pi$$

The most consistent and powerful waves are generated within the belts of the mid-latitude westerlies towards the eastern margins of the major oceans; the UK is particularly favourably placed. In the open Atlantic, some 400 km west of the British Isles, average power levels are around

100 kW m⁻¹ of wave front. This has decreased by half in water depths of 50 m off the Outer Hebrides and is no more than 20 kW m⁻¹ where the depth is 15 m (fig. 7.4). Most practical devices envisage conversion of the waves' mechanical energy to electricity, although overall efficiencies of more than 25% are difficult to achieve partly because of the irregular nature of wave heights, periods and directions encountered in the open ocean. Since the greatest power with which a wave energy conversion device would have to contend could exceed 10 MW m⁻¹, it has proved difficult to find a design sufficiently robust to survive in such a hostile environment long enough to deliver more energy than would be required for its construction, installation and maintenance. Although more than 300 different design proposals have been evaluated in the UK and a number tested up to one-tenth scale, factors such as those mentioned above, together with the added difficulty of transmitting the electrical power generated to distant sources of demand, have indicated that shore-based devices will provide the most cost-effective approach at the present time.

One of the simplest wave energy conversion systems (WECS) is the oscillating water column (OWC) converter. A small-scale version, which can deliver up to 100 W to light buoys, allows passing waves to transmit their periodic motion to a trapped column of air whose movement, through a system of valves, turns an air turbine consistently in one direction. A larger-scale 500 kW version operated on the coast of Norway until it was damaged in a violent storm in 1989. The Japanese have used this technique successfully to produce 600 kW from a research vessel, Kaimai, moored 2 km offshore, although this was considerably less power than had been anticipated.

In the UK, a 75 kW prototype OWC device, using a self-rectifying Wells turbine, is under test on Islay in the Inner Hebrides (fig. 7.5). The construction is built across a natural rock gully and is expected to generate 300 MWh of electricity per year. If successful, a larger 1.5 MW device would be built with the intention of attracting commercial investors. However, the overall potential for this type of installation is limited in the UK by topography and wave climate to about 300 MW.

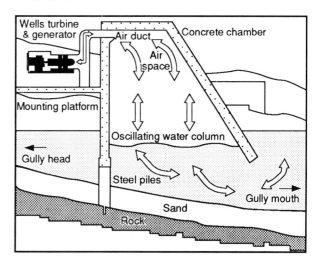

Figure 7.4 **The available wave power (in kW m⁻¹) at selected sites around the British Isles.**
Source: Brown and Skipsey (1986)

Figure 7.5 **The 75 kW prototype wave energy conversion device on Islay.**
Source: ETSU

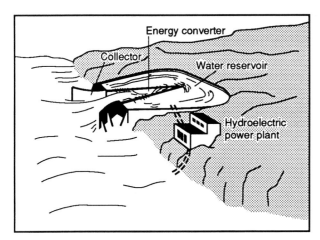

Figure 7.6 **Schematic representation of the Tapchan wave energy conversion device on the Norwegian coast.**
Source: IEA (1987)

A quite different approach involves diffracting an incoming wave front into a progressively narrowing inlet and up a constricting channel such that the wave height is amplified by up to ten times (Tapchan from *Tap*ered *chan*nel). The waves break over a retaining wall, filling a reservoir with water whose head relative to sea-level can be used to drive a conventional low-head turbine. A system of this type, concentrating the energy from a 10 km wave front, operates at the head of a Norwegian fjord (fig. 7.6) and a similar one to deliver up to 20 MW of electrical power has been proposed for Mauritius, which lies within the belt of the south-east trade winds.

The environmental impacts of the small prototype devices so far constructed are quite limited and localized but a large offshore WECS, extracting significant amounts of energy from the waves, would be expected to change the wave climate at the shore with corresponding implications for the movement of sediments and for littoral ecosystems.

Tidal power

Tide mills, which trap water at high tide behind a barrier and release it when the tide is low to provide mechanical power, have operated around the coastline of the British Isles since the Middle Ages. Old London Bridge had four 20-foot reversing paddle wheels mounted below the principal arches which were turned by the force of the fast tidal currents. Modern approaches favour the construction of large-scale barrages to enclose a basin whose filling and emptying generates

electricity from turbines designed to operate on low heads of water.

In the open ocean, the combined gravitational attraction of the Moon and the Sun generates tidal ranges of little more than a metre. They vary predictably in amplitude depending on the relative positions of the Moon and Sun, the rotation of the Earth and the topography of the ocean basins. In the Atlantic Ocean, including the seas around the UK, a semi-diurnal cycle is experienced in which the interval between successive high tides is close to 12 hours 25 minutes; tides in the Pacific are often modified by a strong diurnal component. Tidal ranges along coasts are usually increased by the effects of shoaling, while in enclosed gulfs and estuaries, the combined effects of funnelling and resonant amplification, which depends on particular combinations of water depth and channel length, can lead to ranges occasionally in excess of 10 m (fig. 7.7). Although a total of about 3 TW of tidal power is dissipated worldwide by frictional drag, sites suitable for the harnessing of some of this energy by the construction of tidal barrages are relatively uncommon; the most significant are shown in fig. 7.8. The available power depends on both the tidal range and the size of the impounded basin, whilst turbine efficiency increases with increasing head. Since the size of the basin increases seaward but tidal range decreases, the positioning of a barrage across a bay or estuary requires a degree of compromise.

The simplest type of scheme involves one-way generation of electrical power (fig. 7.9a). Water is

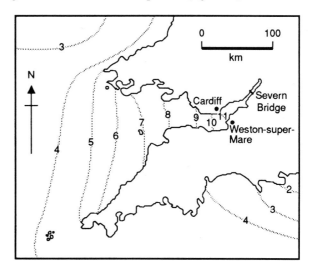

Figure 7.7 **The mean range of spring tides (in m) around the coast of SW England and Wales showing the effect of amplification in the Severn Estuary.**
Source: DEn (1981)

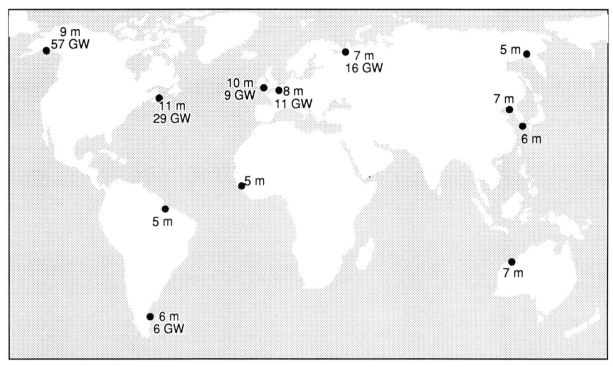

Figure 7.8   **The location of the world's most favourable sites for tidal power generation showing the mean range of spring tides (in m) and, in some cases, the maximum available power.**
Source: Sørensen (1979)

held behind the barrage at high tide and power generation is begun as soon as sufficient head has developed between the impounded waters and the ebbing tide. In this type of scheme, power can be produced for 4 to 5 hours of a 12-hour semi-diurnal tidal cycle. Although intermittent, the periods of power availability are quite precisely predictable well into the future. The benefits of four rather than two daily periods of power production can be gained by installing reversible turbines capable of two-way generation, on both the ebb and flood tides, although this increases engineering costs and does not significantly increase the total power output (fig. 7.9b). At still greater expense and further loss of efficiency, double-basin schemes allow even more flexibility in the supply of power, including the possibility of continuous power generation and use as pumped storage (chapter 8).

An oft-quoted example of a working tidal barrage was completed across the Rance estuary in Brittany in 1966 (fig. 7.10). Here 24 reversible turbines give a peak output of 240 MW and provide France with its cheapest source of electricity. Other smaller schemes are operating in the former USSR, China and Canada. The most promising site in the UK is the Severn Estuary, which has a spring tidal range increasing from 8 m near Ilfracombe to 11 m near Weston-super-Mare

and a feasibility study was carried out as long ago as 1925. The most recent proposal (DEn, 1989; ETSU, 1990) is for a 16 km barrage from Brean Down, near Weston, to Lavernock Point, south of Cardiff (fig. 7.11). It would incorporate 216 turbines each 9 m in diameter rated at 40 MW. The total installed capacity of 8640 MW would provide 17 TWh of electricity per annum, equivalent to 7% of present consumption and an annual saving of 8 Mt of coal. The total cost of construction (in 1988) was given as £8280 million. A detailed feasibility study for the Mersey estuary is in progress: with a more modest 620 MW of installed capacity, it is expected to provide slightly cheaper electricity than the Severn barrage. Many smaller inlets along the western seaboard are considered appropriate for tidal power, and a 30 MW scheme for the Conwy estuary and 9 MW one for Loughor are also being studied (fig. 7.12). It is estimated that the combined output of all the feasible sites (90% from 8 barrages each greater than 500 MW capacity and the remainder from 34 sites each below 100 MW) would supply 20–25% of present electricity demand in the UK.

Large-scale tidal barrages, like OTEC and WECS (wave), require high capital investment and involve lengthy construction periods but, unlike them, are based on relatively familiar and proven technology. There seems no reason why their

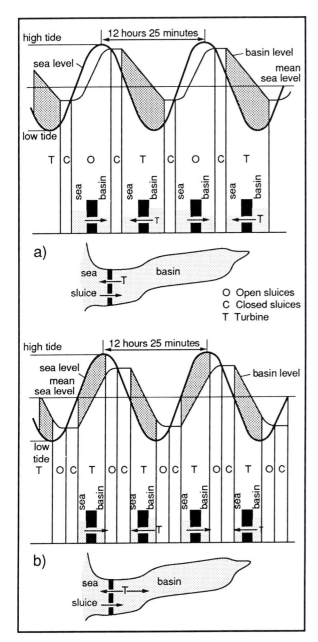

Figure 7.9 The generation of tidal power from a single-basin scheme using (a) one-way (ebb tide) generation and (b) two-way generation.
Source: Brown and Skipsey (1986)

Figure 7.10 The 240 MW tidal barrage across the Rance estuary, Brittany.
Photo: ETSU

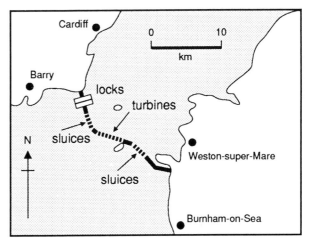

Figure 7.11 A proposed location for a Severn tidal barrage capable of generating 8.6 GW of electrical power.
Source: ETSU (promotional literature)

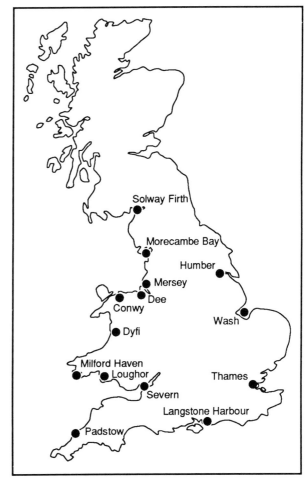

Figure 7.12 Possible locations for tidal barrages around the British Isles.
Source: ETSU (promotional literature)

useful life should not be well in excess of a century. They share the virtues of free fuel, zero pollution and low running costs but they are not without environmental impacts. A preliminary investigation has been completed for the proposed Severn barrage which found that landward of the barrage the lowest tide level would correspond to the present mid-tide and the strength of tidal currents would be reduced. As a consequence, the volume of sediments in suspension, and hence turbidity, would decline while bed stability would increase; all of these factors would in turn have major ecological impacts. There may be adverse effects to inland drainage and water quality as well as benefits such as protection against storm surges and a useful alternative communications link to the Severn Bridge.

Points to consider

- What are the main similarities and differences between the three marine energy sources, together with possible energy conversion systems, which have been described?

- At what scales, if any, are they a practical possibility?

- To what locations are they best suited?

- To what extent is it possible to predict the full range and extent of their impact on the local environment and its inhabitants?

- What contribution, if any, could each one make to the energy needs of the local area and to the nation (UK or other) as a whole?

- How would you rate their acceptability to (i) local residents, (ii) the general public, (iii) tourists, (iv) conservationists, (v) industrialists, (vi) the energy utilities, (vii) the Department of Energy, (viii) the European Community, (ix) the World Bank?

8 Renewable energy on land

- How much energy is available from rivers, the wind and the Earth's internal heat?

- How and where can it be harnessed?

- Are these energy sources entirely renewable?

- What impacts do they have on the environment?

The renewable energy sources considered in this chapter make a rather diverse group: wind energy is collected from the atmosphere above the Earth's surface, hydropower from the surface itself and geothermal energy from beneath it. Conversion devices in each case are most often concerned with the generation of electricity and are most frequently based on land. Compared with the marine sources discussed in the previous chapter, they are more familiar in use and the technology involved is on the whole less unconventional; hence they seem likely to achieve more widespread deployment.

Both wind energy and hydropower derive indirectly from solar energy and are entirely renewable; the source of geothermal energy, contrastingly, is independent of the Sun and it can only rarely be considered renewable within a short time scale. While both hydropower and geothermal energy can provide a firm power source, energy from the wind is intermittent and unpredictable in its availability, although its seasonal variation in temperate latitudes is broadly in phase with demand. Although conversion to electricity is considered to be the most useful

option – at least in MDCs – both hydropower and wind are capable of yielding useful mechanical power, while some geothermal resources are more suited to the supply of low-grade heat.

Hydropower

Mechanical energy has been obtained from flowing water for many centuries, being used for such tasks as grinding corn, draining marshes, irrigating crops and then for driving more complex machinery in the early stages of the Industrial Revolution. More efficient and powerful turbines began to replace water wheels from the mid-nineteenth century, which led on to the generation of electricity in the twentieth.

The gross gravitational potential energy, E_p, dissipated by a river in the course of a year can be estimated as

$$E_p = 0.155 \, l \, s \, Q \text{ TJ}$$

where l is length in km, s the gradient in m km^{-1} and Q the flow rate in m^3 s^{-1} (Mustoe, 1984). Globally, this is equivalent to about 10 TW of power, which is equal to the world annual energy consumption (300 EJ). Although it is possible to utilize the natural flow of a river for the generation of electricity, most hydropower schemes involve the construction of a dam to impound a large reservoir of water in order to secure firmness of supply. The quantity of power that can be developed from a given head of water, h, depends upon the flow rate, Q, according to the equation

$$P = \eta \rho g h Q$$

where η is the efficiency of the conversion process and ρ the density of water.

Modern turbines (fig. 8.1) are capable of converting the gravitational potential energy of the water into electricity with efficiencies close to 90%. There are many different types of design, the choice depending on the available head and flow rate, and the desired power output. They can be broadly categorized as either *impulse*, where a high-velocity jet of water impinges upon a freely spinning multi-bladed turbine (a development from primitive water wheels), and *reaction*, where the turbine is totally immersed in water whose internal pressure imparts a force to the turbine blades (rather like a ship's propeller working in reverse). The former is best suited to high heads and lower flow rates of water, while the latter can operate reasonably efficiently with quite small heads and correspondingly higher flow rates. Since costs generally increase with size of turbine, it is preferable for economic reasons to maximize the head rather than the flow rate.

Hydropower currently supplies just over 20% of the world's electricity. It has been estimated (WEC, 1989) that 70 EJ a^{-1} (2.2 TW) represents a reasonable potential for hydropower, of which only 17% has been harnessed. While the MDCs have realized almost half of their potential, the corresponding figure for the LDCs is only 7%, yet this represents almost half of their total electricity production. The potential for hydropower tends to be greatest where topography is most rugged and rainfall highest; such regions tend to be remote from regions of greatest demand and are often areas of wilderness or outstanding natural beauty. In MDCs most of the obvious sites have already been exploited; in LDCs a number of schemes in the 5–25 GW range have been proposed, and a few implemented, but it is not always clear how such large quantities of electrical power will be utilized or who will be the beneficiaries.

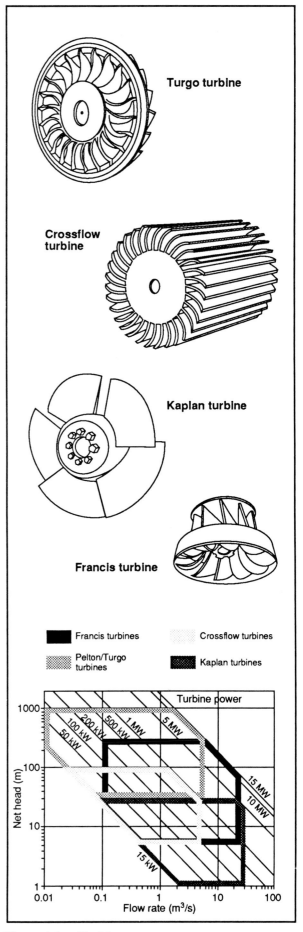

Figure 8.1   Turbines
Source: ETSU

Large-scale hydropower schemes share many features in common with large tidal barrages: the technology is well-understood, running costs are low, the fuel is free and constantly available, and its conversion creates no pollution. They are, however, very capital-intensive and, although long-lived in principle, their productive lives can be reduced to a few decades by a build-up of sediment behind the dam. Benefits additional to the generation of electricity can often be available, including flood control, irrigation, aquaculture, and opportunities for recreation and tourism, but these must be balanced against the possible consequences of reduced river flows and suspended sediment loads below the dam and the flooding of large tracts of land above it. The latter may involve the loss of important habitats, the displacement of wildlife and indigenous populations (up to 100,000 for the largest schemes) and the exacerbation of soil erosion. In warmer climates, the quality of the impounded water can deteriorate to the extent that it becomes unsuitable for supplying drinking or irrigation water and facilitates the spread of water-borne diseases.

The UK obtains rather less than 2% of its electricity from hydropower (generated mostly in Scotland) with little scope for expansion of installations feeding significant power into the national grid. An alternative approach, which is applicable worldwide and avoids many of the drawbacks of large-scale developments, involves

small-scale 'run-of-river' devices employing weirs or very small dams. A survey carried out in Wales (E.M. Wilson in WCE, 1985), for example, conservatively identified more than 500 potential sites for generation above 25 kW; the very many smaller sites, of a capacity appropriate to single farmsteads, for example, were excluded from the review. A survey in the Dyfi Valley (Ashby, 1979) similarly found over 200 sites where mechanical power had formerly been generated but which might now yield up to 6 MW of electricity. A nationwide survey has been in progress since 1987 and expects to identify a total potential installed capacity in the region of 500 MW. The efficient utilization of heads below about 3 m will, however, require the development of innovative technology based, for example, on hydropneumatic principles.

Amongst LDCs, China is one of a number which exemplifies a policy of 'walking on both legs': it has a continuing programme of some 100,000 mini-hydropower installations, whose average size is about 80 kW, but at the same time is developing multi-gigawatt barrages on its major rivers.

Mention should finally be made of the application of the technology to *pumped-storage hydropower* systems, most impressively at Dinorwig in Snowdonia (fig. 8.2). Here advantage has been taken of a natural mountain lake nearly 600 m above an abandoned quarry at the level of the generating equipment. Within a period of six

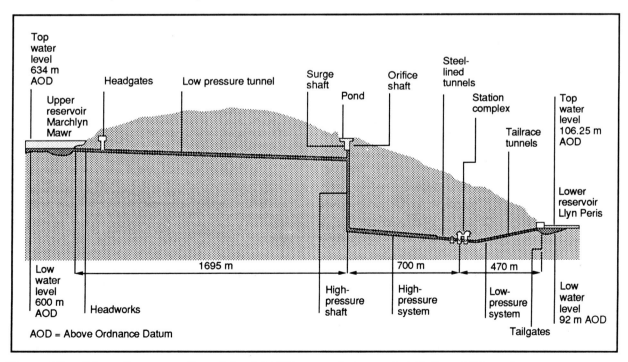

Figure 8.2   Schematic cross-section of the 1.5 GW pumped storage installation at Dinorwig.
Source: ETSU

hours, six 300 MW reversible turbines are able to pump more than 7 Mm³ water from the quarry level to fill the high-level lake. When needed, the water descends, at a maximum rate of 420 m³ s⁻¹, to the lower level through the turbines whose peak output of electrical power is available for a period of five hours within ten seconds of initiating the flow. This type of installation allows electricity (especially that generated by nuclear stations, which provide baseload power since they do not respond well to varying output) to be stored at times of slack demand and returned to the grid whenever demand suddenly increases. The efficiency is as much as 80%, but this figure does not take into account the considerable losses incurred in the generation of the stored energy from primary fuels.

Power from the wind

Winds blow over the Earth's surface in response to the uneven distributions of temperature, density and pressure in the atmosphere which result from non-uniform solar heating. The actual flow is considerably modified by the effects of the Earth's rotation as well as its surface topography. While on a global scale winds blow in well-recognized patterns, such as the trade winds and the mid-latitude westerlies, at smaller spatial and temporal scales there will be frequent variations in wind speed and direction, sometimes, as in the case of sea breezes, following a diurnal cycle.

Like water power, wind power has a long history of use for driving simple machinery and pumps. Where the requirement is for high mechanical power (torque), blades rotated by frictional drag and whose tip speed is rather below that of the wind are appropriate. This includes types such as the familiar windmill of the UK and Holland, and the multi-bladed wind pumps found commonly in rural areas of Australia and the USA. For the generation of electricity, however, higher ratios of blade tip speed to wind velocity are more efficient; devices that obtain rotational energy by aerodynamic lift resemble aircraft propellers and are usually referred to as *aerogenerators* or *wind turbines*. A general term applied to all types of device is *wind energy conversion system* or *WECS* (not to be confused with wave energy conversion system) and a few characteristic examples are sketched in fig. 8.3.

The power of the wind P_w is given by

$$P_w = \frac{1}{2} \rho A v^3$$

where ρ is the density of air, A the area swept by the blades and v the wind speed. Thus the power obtained increases as the square of the blade diameter and as the cube of the wind speed. In theory an aerogenerator is able to extract almost 60% of this power but practical devices rarely have conversion efficiencies much above 40% and such values are achieved only within a range of design windspeeds. Whilst the most common design of wind turbine employs a horizontal axis rotor mounted with the generator at the top of a tall tower, vertical axis machines are feasible and have the advantage that they respond equally well to winds from any direction and can be of lighter construction since the gearbox and generator are at ground level (fig. 8.4). For all types of design, care must be taken to guard against blade fatigue caused by the high and variable stresses.

For a typical horizontal axis turbine, power output rises rapidly from a cut-in windspeed, below which electrical generation is not possible, up to a maximum rated value, maintained up to the point at which structural damage might be sustained (fig. 8.5). At this cut-out windspeed the blades may be feathered or otherwise protected from self-destruction. 'Off-the-shelf' turbines are now available up to about 750 kW rated output and there are a number of prototype designs in the 1–5 MW range. Although wind power inevitably suffers from variability and unpredictability of supply, it can be deployed at scales from a few kW to MW, whilst larger outputs of power can be built up incrementally from arrays of individual turbines. Such developments, making effective use of the most suitable sites, are known as *windfarms*, where the spacing of the individual turbines is determined by the need to limit mutual interference of the airflow (which increases the stress on the blades).

The dependence of power output on the cube of the windspeed suggests that much is to be gained from siting wind turbines at locations having the highest mean annual windspeeds together with a high percentage duration of periods of strong winds. In general, suitable locations are to be found in more remote areas such as on hilltops and along exposed coasts, which may also be areas of outstanding natural beauty and far from centres of electricity demand. In such cases, a wind turbine of modest scale might serve an isolated local community whilst larger installations could feed

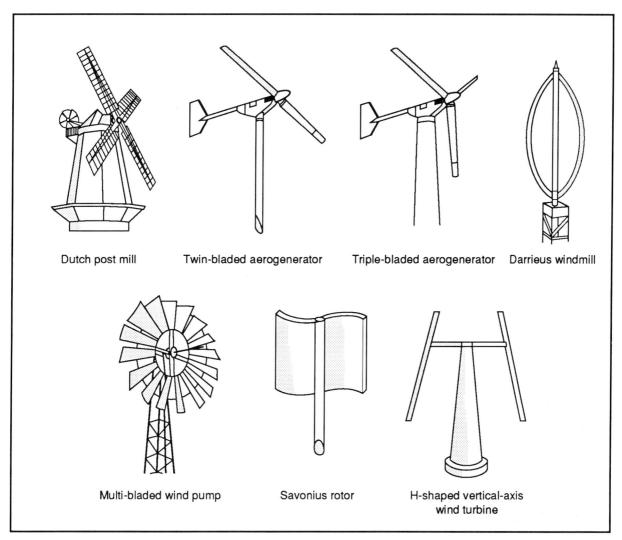

Dutch post mill   Twin-bladed aerogenerator   Triple-bladed aerogenerator   Darrieus windmill

Multi-bladed wind pump   Savonius rotor   H-shaped vertical-axis
wind turbine

Figure 8.3   Typical designs of wind energy converters. The Savonius rotor has become familiar from its adaptation for use in free-standing kerb-side advertising signs and for roof-mounted ventilators on commercial and public-service vehicles.

power into a national grid. It has been proposed that offshore deployment could help to avoid aesthetic intrusion and competition for the use of land, whilst at the same time benefiting from the higher velocity and less turbulent windflow over the sea.

One of the earliest and largest developments of windfarms was at the Altamont Pass some 50 km east of San Francisco Bay. At about 250 m up in the Coast Range, the pass experiences consistently strong winds with a diurnal cycle. A favourable early-evening peak intensity results from the considerable temperature contrasts that develop between the cool Pacific Ocean to the west of the Coast Range and the San Joaquin Valley on its landward side. The strength of the winds is further intensified by the funnelling effects of the pass. Development here was also encouraged by the relatively unproductive nature of the land and a system of state and federal tax credits which

attracted the interest of investors. In California as a whole there are almost 18,000 wind turbines, mostly in the 50–300 kW power range with a total rated output of about 1.5 GW; it is anticipated that by the year 2000 wind turbines will supply 8% of California's demand for electricity.

Although the total amount of power in the atmospheric circulation is of the order of 10 PW, only a very small fraction of this can be extracted by WECS. An estimate of the distribution of world resources is given in fig. 8.6, from which it can be seen that the UK is very favourably placed. It is generally agreed that significant power generation requires annual average hourly mean windspeeds (measured at 10 m above the ground) greater than about 5 ms^{-1}, although mean speeds as low as 2.5 ms^{-1} are adequate for small machines. For the UK the figure of 5 ms^{-1} is exceeded around the coasts and inland in the west and north; mean windspeeds on hilltops are naturally higher still

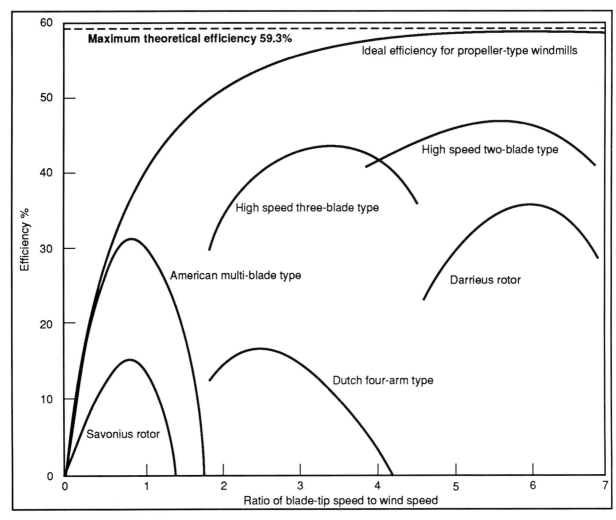

Figure 8.4   The efficiencies of different types of wind energy converter.
Source: BWEA (1982)

Figure 8.5   The variation of power output with windspeed for a typical aerogenerator.
Source: Kovarik *et al.* (1979)

and more than 3000 such locations have been identified with mean speeds above 8.5 ms⁻¹.

A number of wind turbines have been constructed in the UK which are supplying power to the grid (fig. 8.7), the largest being the 3 MW, two-bladed, 60 m diameter aerogenerator installed on Burgar Hill, Orkney, in 1987. The site at Carmarthen Bay includes a 130 kW prototype variable geometry vertical axis wind turbine (VAWT) and a recently commissioned similar 500 kW model intended for use in windfarms. Beginning with a site at Capel Cynon, Dyfed, it is planned to develop up to three windfarms each having about 25 turbines at 300–400 m spacing (about 10 rotor diameters) with a total power output around 8 MW.

The environmental impact of wind turbines relates chiefly to their visual intrusiveness in the landscape, to the propagation of low frequency noise and to interference with telecommunications, especially television reception; some concern has also been expressed regarding the risks to birds and aircraft. Investigations are continuing in the UK into the public acceptability of both single turbines and the possibility of windfarms; it is considered that offshore development is at present not an economic proposition.

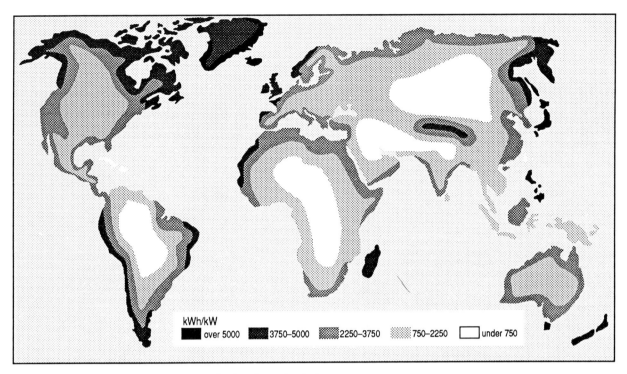

Figure 8.6   Estimated world-wide distribution of wind energy sources.
Source: IEA (1987)

Figure 8.7   Wind turbines at Carmarthen Bay, Wales. Left to right: VAWT (500 kW); HAWT (300 kW); VAWT (130 kW); HAWT (300 kW).
Photo: National Power

Geothermal energy

Temperatures within most of the Earth's interior exceed 1000°C, reaching about 4000°C at the centre of the core. Some of this heat is primordial but the rate of cooling over the past 4500 Ma has been slowed by the continuing decay of radioactive elements, principally uranium-235 and -238, thorium-232 and potassium-40. Since the surface temperature of the Earth is about 15°C, it is to be expected that temperatures would increase quite rapidly with depth in the crust and an average temperature gradient of about 30 K km^{-1} is found, although it can exceed 100 K km^{-1} in geologically active regions.

Heat is being conducted to the surface of the Earth from the interior according to the equation

$$q = K \, \Delta T/\Delta z$$

where q is the heat-flow per unit area, K is the average thermal conductivity of crustal rocks and $\Delta T/\Delta z$ is the temperature gradient. The global average value is close to 60 mW m^{-2} although it may be more than an order of magnitude greater in volcanically active areas. The total heat-flow through the continents is therefore some 50 TW; though large in magnitude, it is too diffuse a resource to be of practical value. An alternative view (WEC, 1980) notes that the heat content of the uppermost 3 km of the continental crust is some 40 million EJ, which, if the technology were available for 'heat mining', represents an even larger resource, although it would be non-renewable except over long time-scales.

It follows from the above equation that, given a steady flow of heat, higher than average temperatures may be encountered at relatively shallow depths in sedimentary basins incorporating strata of low conductivtity. The same is true for bodies of igneous rock containing unusually high concentrations of radioactive isotopes. However, if useful quantities of heat are extracted from these types of source it is replaced at a much slower rate; hence they are not renewable but constitute a form of heat mining. At certain locations adjacent to a boundary

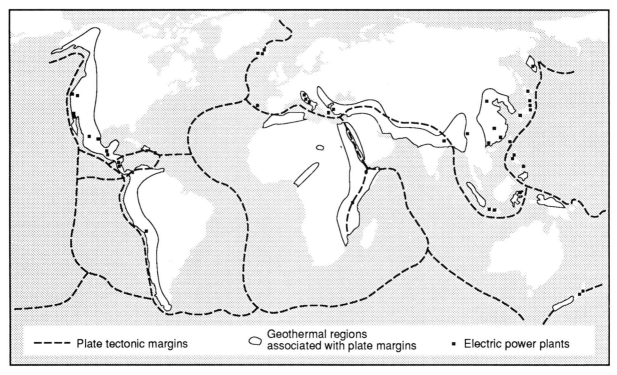

Figure 8.8   The location of the world's high-temperature geothermal fields.
Source: IEA (1987)

between tectonic plates, volcanic activity results in very much elevated temperatures and the flow of heat to the surface is sufficient to match the rate of removal; these less common occurrences therefore constitute a renewable resource.

The surface of the Earth may be classified geothermally into three types of region: *hyperthermal (high enthalpy)*, where the temperature gradient exceeds 80 K km^{-1}; *semithermal (low enthalpy)*, where it lies between 40 and 80 K km^{-1}; and *non-thermal (normal)*, where it falls below 40 K km^{-1}. Natural hydrothermal circulations can extract useful quantities of heat from the first two types while artificial fracturing of igneous bodies within the second type enables water pumped through the rock to extract the heat. These three principal kinds of geothermal energy resource will be considered in more detail.

Hyperthermal fields
With high temperatures and, at least in the short term, an inexhaustible supply of heat, hyperthermal fields are the most useful and most widely exploited types of geothermal resource despite their restricted occurrence (fig. 8.8). An essential requirement is for the presence of an intrusive body of magma within about 10 km of the surface to provide a sustained source of heat which can raise the temperature of aquifers at shallower depths. It is important that the aquifers

can be readily recharged from precipitation and that impervious strata overlying the aquifer(s) prevent the easy escape of the superheated water or steam generated by the high heat-flow and pressure. (The required configuration bears some resemblance to that for hydrocarbon traps described in chapter 3.) Hot springs and geysers frequently betray the presence of this type of resource.

By drilling down to the aquifer, in the most productive *dry fields*, steam at temperatures exceeding 200°C can be brought to the surface at pressures sufficient to drive turbines for the direct generation of electricity. Elsewhere, in *wet fields*, superheated water (accompanied by some steam) is flashed into steam after reaching the surface, although this can lead to an unacceptably high build-up of scale from precipitation of dissolved salts. For this reason, or where water temperatures below about 200°C reduce conversion efficiencies too far, it is preferable to evaporate a working fluid of lower boiling point (isobutane, for example) in a *binary cycle* plant (similar in principle to OTEC, chapter 7).

Potential sources of pollution arise from the steam, which may be accompanied by noxious gases, particularly hydrogen sulphide, and from the condensed water, which contains high concentrations of salts leached from the aquifer. Although the hydrogen sulphide may be removed

Table 8.1   Use of geothermal power for electricity generation and directly as heat

| Country | Direct heat | | Electricity | | |
	Installed capacity MW$_{th}$ (thermal)	Principal type of application in order of significance	Installed capacity MW$_e$ (electrical)	Planned capacity in AD2000	Total equivalent energy produced (ktoe per year)
China	35	A, H, I, B	18	32	241
El Salvador			95	175	133
France	676	H, A			266
Hungary	1580	B, A, X, H, I			630
Iceland	1306	H, B, A, I	41	71	708
Indonesia	small		88	1272	133
Italy	631	B, H, A, I	548	1400	1014
Japan	4764	B, X, A, H, I	215	353	1978
Kenya	small		45	120	63
Mexico	small		700	1290	990
New Zealand	176	I, H, B, A, X	167	317	346
Nicaragua			70	180	124
Philippines	small		894	2266	1260
Romania	273	B, A, H, X	small		96
Turkey	200	B, X, H, A	21	130	105
USA	1776	H, I, B, A	2212	3331	3760
former USSR	1404	H, A, M, I	11	241	656
Other	808		11	520	314
Total	13,629		5136	11,698	12,817

A – agribusiness, aquaculture   H – space heating/cooling, hot water   I – Industrial process heat
B – medicinal and recreational bathing   X – undefined

Source: IEA (1987); WEC (1989)

by scrubbing or conversion to useful elemental sulphur, it is unfortunate that its characteristically unpleasant odour can be detected at concentrations down to 2 ppb by volume, well below the toxic threshold. As an alternative to discharge to surface drainage, the condensed water may be reinjected deep underground but at a point sufficiently distant from the extraction well(s). At certain locations, notably the Wairakei field, New Zealand, land subsidence – here up to 40 cm per annum – has been caused by the withdrawal of fluids from underground; reinjection seems not to provide a complete solution. It has also been anticipated that the withdrawal and reinjection of fluids may induce or increase seismic activity in geothermal areas, but it has been demonstrated that sufficiently low injection pressures can effectively eliminate this problem.

Geothermal power stations, which typically provide electrical power outputs in the 10 to 100 MW range, are no different from conventional ones in that their efficiency is limited by thermodynamics. Where a demand exists, an alternative, more efficient use for the hot water or steam is as district heating or process heat for domestic, commercial or industrial consumers. Such a use involves passing the water or steam through heat exchangers. The present and projected future extent of high-grade geothermal sources for both direct heating and electricity generation are shown in table 8.1. Although of limited extent globally, the resource nevertheless has considerable local significance. In Iceland, for example, more than 80% of houses have geothermal space heating, while in California as much as 2 GW of electrical power is geothermally generated. Geothermal heat also has the potential to be a significant source of power for many developing countries, such as Chile, Colombia, Costa Rica, Ethiopia, Guatemala, Peru and Tanzania.

Aquifers in semithermal fields

Though less useful than the high-grade sources discussed above, hot groundwaters at temperatures below 100°C are widely available in sedimentary basins where the presence of poorly conducting impervious strata, such as shales, mudstones, coal seams or evaporites, produces higher than average temperatures (at a given depth) in intercalated water-bearing sandstones or limestones.

One such development occurs in the Paris Basin, where groundwater, which reaches temperatures of 50 to 75°C at a depth of 1.3 to 1.9 km in Middle Jurassic sandstone aquifers, has been utilized in a number of district heating schemes on the outskirts of Paris. A typical installation, as at Mélun, provides hot water and space heating for about 1000 purpose-built high-density apartments from about a dozen wells, with individual flow rates of about 80 litres per minute on a 1 km grid. Pairs of wells, for extraction and reinjection, are drilled at a high angle, a technique known as *whipstocking*. It avoids 'short-circuiting' the warm water flow while at the same time maintaining underground pressures and disposing of saline reject water (typically containing 25 g per litre of salts). The system supplies about 70% of annual heating requirements but the life of individual wells is expected to be limited to 30 or 40 years.

A number of comparably deep sedimentary basins occur in the UK (fig. 8.9) but it is not anticipated at the present time that they will provide economically useful quantities of heat. However, a small district heating scheme is operating in a new development in Southampton city centre from an aquifer in the underlying Hampshire Basin, whose heat output is limited by the modest fluid flow rates realized.

Hot dry rock

A third type of geothermal resource occurs in large shallow intrusions of granitic rock where above-average concentrations of radioactive isotopes result in geothermal gradients of 60 to 70 K km^{-1}. Hence by drilling to relatively moderate depths, it is possible to obtain temperatures above 200°C. However, since such rock bodies are not natural aquifers, it is necessary to induce or intensify permeability within a volume of the rock and then to extract its heat content by pumping water down from the surface (fig. 8.10). The approach involves drilling an injection well which curves to make an angle of about 60° with the vertical; a small explosive charge is detonated at the bottom to improve hydraulic contact between the borehole

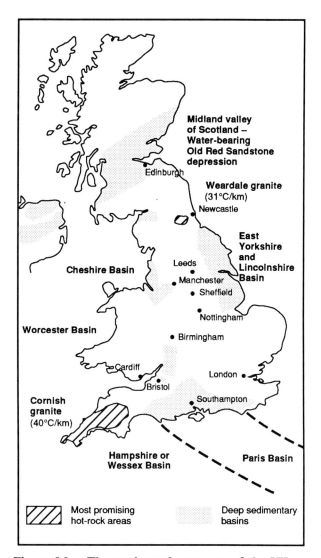

Figure 8.9   The geothermal resources of the UK.
Source: Flood (1983)

and the natural system of joints and fractures. The natural network is then opened up by pumping down water, or higher-viscosity fluid, at very high pressure (a technique known as *hydrofracturing*) in order to create a large surface area for heat transfer. Finally, connection is made within the fractured zone with the production well, also angled, to complete the hydraulic circuit. With sufficiently high temperatures at depth, injected water will return as steam and can be used directly for generation of electricity or through a binary cycle if temperatures are too low.

Experiments early in the 1970s at Fenton Hill, New Mexico, by the Los Alamos National Laboratory demonstrated the feasibility of the concept and this work has progressed to the point of producing 9 MW (thermal) of energy with an exit temperature of 190°C. Work was begun only a little later by the Camborne School of Mines at Rosemanowes Quarry, Cornwall (fig. 8.11), on an

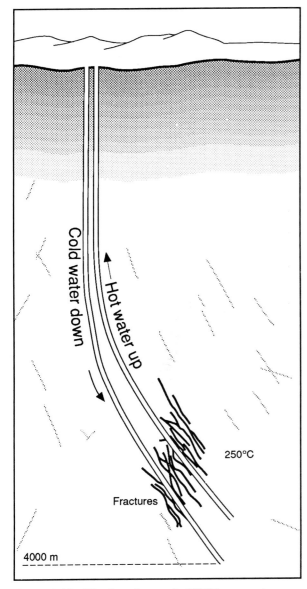

Figure 8.10 The hot dry rock (HDR) concept.
Source: IEA (1987)

Cold water down

Hot water up

250°C

Fractures

4000 m

Figure 8.11 View of Camborne School of Mines geothermal project at Rosemanowes Quarry, Cornwall.
Photo: ETSU

total resources of the Cornish granite batholiths have been estimated as some 200,000 PJ (the equivalent of 8 Gtce), it appears that commercial exploitation remains a long-term prospect.

Points to consider

- How would you rate the relative merits and drawbacks of the three land-based energy sources which have been discussed?

- At what scales and in what parts of the world is hydropower a valuable resource?

- How serious is the environmental impact of large hydropower installations?

- Is wind power a particularly appropriate option for the UK?

- How acceptable do you consider windfarms, either on land or offshore, are likely to be to the general public?

- To what extent is geothermal energy a renewable resource?

- What are the main limitations in using geothermal energy as a source of electricity or heat?

exposed surface of the Carnmenellis granite batholith which has two intersecting sets of prominent joints. Three boreholes have been drilled and a fractured volume of rock created at a depth of 2.5 km. But it has been found necessary to improve the hydraulic performance of the reservoir and to counteract the effects of short circuits before proceeding to a pilot plant for commercial electricity generation which would involve depths of about 6 km, where relatively little is known of rock properties. Although the

9 The fifth fuel – energy conservation

- Must global energy consumption continue to rise?

- What savings are possible in homes, public buildings, shops, offices and factories, and by transport?

- How can reject heat from power stations be used for space and water heating in urban areas?

- Do significant energy savings imply a fundamental review of life-styles which are predicated upon continually increasing production, consumption and resource exploitation?

The relative importance of different sources of primary energy for OECD countries, the (former) centrally planned economies and the UK are shown in fig. 9.1. It has already been pointed out in chapter 2 that not only do fossil fuels account for the greater part of primary energy consumption in most MDCs, but that, as a consequence of the physical laws relating to energy conversion, more than half of this energy input performs no useful task and is discharged to the environment as low-grade heat. It is necessary, therefore, to consider the extent to which it is possible, in both practical and economic senses, to use energy more efficiently, thereby extending the life of non-renewable fossil fuels and allowing a longer time in which to develop alternative supplies.

The domestic sector

In most MDCs, energy consumption is approximately equally divided amongst the domestic and commercial, the industrial, and the transport sectors, each one accounting for between a quarter and two-fifths of delivered energy (fig. 9.2). Whilst there is scope for significant reductions in energy use in all sectors, it is generally considered that in the UK the greatest potential lies in the domestic sector. Energy conservation in both public and private buildings is important if only because the building stock has a much longer useful life than either industrial plant and machinery or road vehicles and aircraft. On average about 85% of domestic energy consumption in the UK is devoted to space and

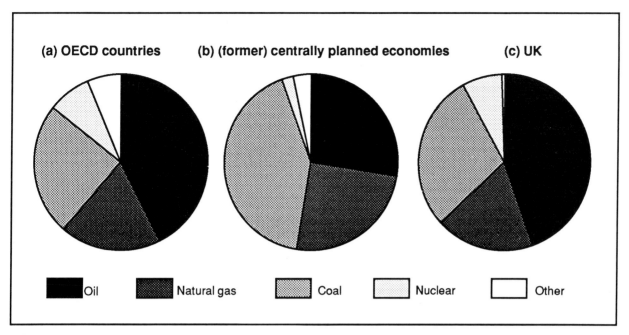

Figure 9.1   Primary energy requirements by fuel for (a) OECD countries (b) (former) centrally planned economies (c) the UK.
Source: IEA (1989c); DEn (1990a)

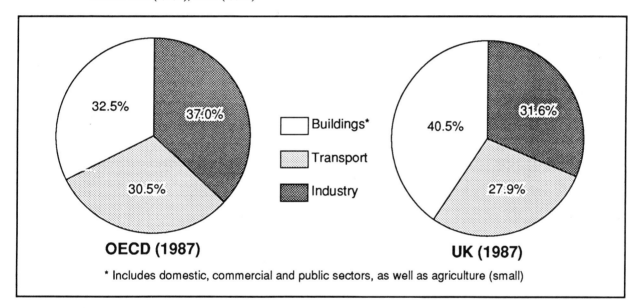

Figure 9.2   The use of energy by sector in (a) OECD countries and (b) the UK.
Source: IEA (1989a), DEn (1990a)

water heating (fig. 9.3), so it is here that conservation measures should be most effective. An older dwelling built without regard for energy conservation can lose as much as 85% of internal heat through doors, windows, walls, floors and roof (fig. 9.4). Raising the temperature of the indoor environment by 1°C can increase heat demand by up to 10%.

Two complementary approaches to the reduction of domestic energy consumption may be distinguished. The first involves economizing on the use of energy by lowering indoor temperatures and dressing more warmly to maintain comfort. The second approach aims to avoid unnecessary use of heat without necessarily compromising standards of comfort; appropriate actions include improving control of heating and lighting in unoccupied rooms, using more efficient domestic appliances, and decreasing heat losses through the fabric of the building and excessive ventilation. Both of these approaches could of course be combined with the passive and active uses of solar radiation discussed in chapter 6.

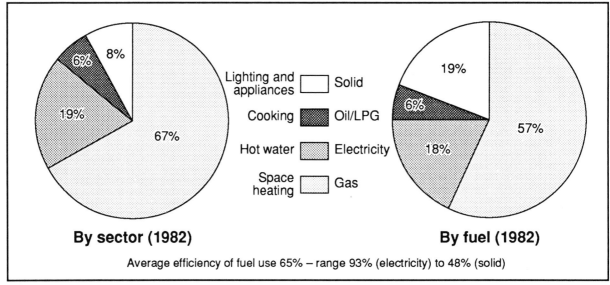

Figure 9.3 Domestic energy consumption in the UK.
Source: DEn (1990b)

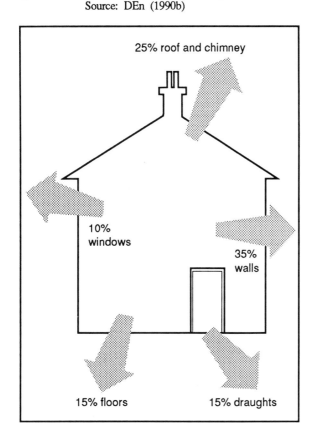

Figure 9.4   Heat losses from a typical semi-detached house built in the 1930s and having full gas-fired central heating.

It has been estimated (DEn, 1990b) that realistic improvements in efficiency of 10% are possible for cookers, 20% for television sets and washing machines, and 50% for refrigerators. The adoption of miniature fluorescent bulbs for lighting would reduce consumption by 75%, whilst gas condensing boilers for central heating achieve savings of 15 to 20%. In theory, heat pumps which work on the same principles as a refrigerator, extracting heat from a low-temperature source (such as the ambient atmosphere or ground outside) and delivering it to the indoor environment, are very efficient devices; however, their widespread adoption has been hindered by high capital costs and uncertainties about their reliability, life-expectancy and maintenance costs (Reay and McMichael, 1988).

The insulation of the housing stock is perhaps the most promising area for energy conservation. The measures required, roughly in order of decreasing cost-effectiveness and increasing complexity of implementation, include draught stripping, insulation of lofts and hot-water tanks, cavity (and ultimately, solid) wall insulation, and double glazing (fig. 9.5). It is suggested (DEn, 1990b) that as much as 50% of the delivered energy used for space and water heating could realistically be saved by the adoption of all the measures outlined.

Whilst it is technically possible to design houses that require negligible amounts of supplementary heating, capital costs are inevitably higher (fig. 9.6). Special attention must be paid, for example, to providing ventilation that maintains a wholesome environment but also conserves energy. At the same time, the life-styles of the occupants may be unacceptably altered. An important advantage of the approaches outlined above is that they can be applied to the existing housing stock and can be implemented incrementally on a DIY basis. It has also been pointed out that a nationwide programme of house insulation, which would include those householders unable to carry out their own, is labour-intensive and could significantly increase employment opportunities.

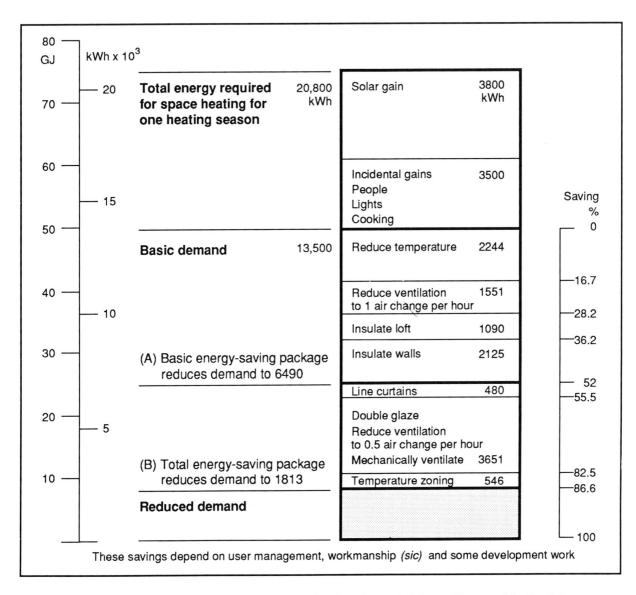

Total energy required for space heating for one heating season	20,800 kWh	Solar gain	3800 kWh
		Incidental gains People Lights Cooking	3500
Basic demand	13,500	Reduce temperature	2244
		Reduce ventilation to 1 air change per hour	1551
		Insulate loft	1090
(A) Basic energy-saving package reduces demand to 6490		Insulate walls	2125
		Line curtains	480
(B) Total energy-saving package reduces demand to 1813		Double glaze Reduce ventilation to 0.5 air change per hour Mechanically ventilate	3651
		Temperature zoning	546
Reduced demand			

Saving %
0
16.7
28.2
36.2
52
55.5
82.5
86.6
100

These savings depend on user management, workmanship *(sic)* and some development work

Figure 9.5 **Possible savings in energy for space heating for the typical house illustrated in fig. 9.4.**
Source: Agius (1979)

Figure 9.6 **The energy-saving house at Machynlleth, Powys.** This house, built by Wates Homes in 1976, was designed to minimize heat losses. It has 460 mm of insulation to all exterior surfaces, quadruply glazed windows, controlled ventilation with heat reclamation and a heat pump. As a result, energy consumption is reduced to about 7% of that in the equivalent conventional house. Note that comparable energy savings may be achieved with lower levels of insulation combined with passive and active solar gain measures.
Photo: Centre for Alternative Technology

Although it is possible to legislate through the Building Regulations for increased standards of insulation in new dwellings and other buildings, the problem of raising the insulation standards of existing (especially older) houses must be addressed through education and persuasion, including advertising campaigns and systems of grants or tax rebates. In the USA, a number of energy utilities have found it more cost-effective to encourage their consumers to raise insulation standards, through measures such as the provision

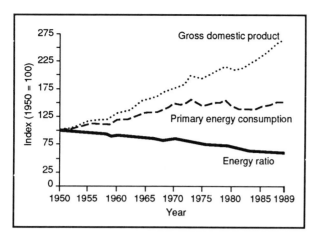

Figure 9.7 **Gross domestic product, primary energy consumption and energy ratio for the UK from 1950 to 1989.**
Source: DEn (1990a)

of free energy audits and zero or low-interest improvement loans, than to invest in new electrical generation plant. It is in such ways that conservation comes to be regarded as the 'fifth fuel'.

Industry and transport

Greater progress in improving energy efficiency has been made in industry during the last two decades, prompted by the rising costs of primary energy, increasingly competitive markets and rising interest rates. In the UK the trend has been encouraged by the establishment of the Energy Efficiency Office (EEO) with regionally deployed efficiency officers. Many firms have benefited from the appointment of an energy manager. A particularly fruitful area for making energy savings in a wide range of industries lies in heat recovery from process heat and there have been considerable benefits from the application of microprocessor control to all forms of energy supply. While world industrial production declined by 9% over a ten-year period from 1973, energy consumption fell by 31%. Similarly, the energy ratio of the UK, defined as the ratio of total primary energy consumption to gross domestic product (at constant prices), fell by about 1% per annum between 1950 and 1973, and by about 2% per annum thereafter (fig. 9.7.).

The recovery of energy in different ways from organic waste products of various kinds was described in chapter 6. It was mentioned that as a preliminary to energy recovery from municipal refuse, metal and glass wastes should be removed.

As well as reducing the demand for raw materials and stresses on the environment, the recycling of scrap metals and glass cullet (and indeed plastics and paper) can result in significant savings in energy. This may be as much as 95% for aluminium, 65% for iron and 25% for glass (Miller, 1990; Porritt, 1990a). The possible advantages to industrial concerns of generating their own electrical power and process heat will be considered below.

The internal combustion engine has an efficiency not much above 10% and it is the private motorist who accounts for about half of the total fuel used in the transport sector. Before the 1973 oil crisis, fuel consumption was not a very significant factor in car design, but the increasing cost of petrol has encouraged motor manufacturers to pay more attention to this aspect. It is also possible to increase fuel savings by other measures, such as encouraging the use of public transport, avoiding unnecessary driver-only journeys through car pooling arrangements and lowering speed limits on trunk roads. Similar considerations apply to freight transport: it is estimated that railways consume two to three times less fuel for the equivalent load and distribution compared with road transport; for water the margin is even greater at about twenty times.

Combined heat and power (CHP)

It has been pointed out more than once that the conversion of fossil fuels into electricity in power stations is necessarily an inefficient process. The most modern coal-fired stations attain efficiencies of almost 40%, although the average for the UK is about 34%. The fate of most of the energy content of the primary fuel is to be discharged to the environment as low-grade heat, mainly in cooling towers. The temperature of the steam when it is finally expelled from the low-pressure turbine at the end of the generation cycle is deliberately made as low as possible (30–35°C) in order to extract as much useful work from it as practicable and to maximize thermodynamic efficiency which, according to the second law of thermodynamics, is directly related to the difference between the initial and final temperatures of the steam. The maximum theoretical efficiency achievable is given by

$$\eta = (T_h - T_c)/T_h$$

where T_h and T_c are the initial and final

temperatures respectively (in degrees Kelvin) of the steam. Since metallurgical limitations impose an upper limit to temperature of about 900 K while ambient temperatures impose a lower limit of about 300 K, efficiencies cannot much exceed 65% in theory and are made even lower in practice by unavoidable losses within the cycle. At around 35°C the temperature of the final cooling water is too low to be of much practical value, although a few small-scale schemes involving fish farming and glasshouse heating are in operation.

An alternative procedure is to terminate the power generation cycle at an earlier point where the steam still retains a useful heat content, at a temperature of about 150°C, even though this reduces the efficiency for electricity generation. The hot steam can then be passed through heat exchangers and its energy content transferred to the circulation pipes of a district heating network to provide space and water heating for domestic and commercial users or process heat for industrial consumers. In this way, as the hypothetical example in fig. 9.8 demonstrates, although the efficiency for electricity generation is always reduced, the overall efficiency for utilization of the energy content of the primary fuel will be more than doubled. This type of installation is known as combined heat and power (CHP) and is relatively common over much of north-west Europe, where power supply is usually the responsibility of the local authority, but is rare in the UK.

In the example considered, the heat-to-power ratio (power supplied as heat) / (power supplied as electricity)

$$= 100 \text{ W} / 40 \text{ W}$$
$$= 2.5$$

which is a typical annual average value for domestic consumers. However, the ratio varies from a winter maximum of about 4 to a summer minimum of about 0.5, and there will be daily and hourly fluctuations as well. For CHP applications it is preferable to use an intermediate pressure turbine which is capable of a degree of variation in the ratio of heat to power delivered. Small surpluses (or deficits) of electricity may be sold to (bought from) the local grid and it is possible to store small surpluses of heat on an hourly basis. In most cases, a CHP system would meet the increased demand for heat in the coldest weather by augmenting output with a heat-only boiler.

Because it is both costly and inefficient to transmit heat over any great distance, CHP systems must in

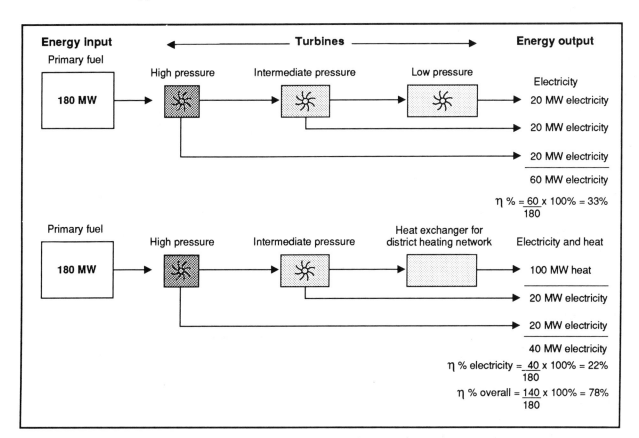

Figure 9.8   Illustrative comparison between a small conventional power station producing electricity and the same station adapted to CHP.

practice be based on relatively small power stations located within or very close to urban centres. Furthermore, the costs of installation are reduced if the heat is supplied to high-density housing developments where the necessary heating mains can be laid during the phase of construction. The laying of heating mains in already built-up areas, especially in low-density residential areas, is more costly, time-consuming and disruptive. Recent policy in the UK has been to build very large power stations, typically about 2 GW output, in semi-rural sites away from large centres of population, though often close to major coalfields. This economy of scale maximizes generation efficiencies whilst the tall multi-flue stacks reduce the impact of local pollution (fig. 3.3). Nuclear power stations also have relatively large power outputs and are placed in even more remote locations, usually on the coast. The present generation of UK power stations is, therefore, poorly suited for CHP applications.

In order to maximize the overall benefit, it is usual for all consumers within the area of a CHP scheme to be connected to the system, but this clearly limits their individual choice of preferred fuel for heating and cooking. The additional expense of installing a natural gas supply is unlikely to be contemplated. The monitoring of the heat supplied to consumers involves measuring (and then obtaining the product of) both the flow rate and the difference in the inlet and outlet temperatures of the circulating water (or steam). Such meters are more complex in design, more costly and require more maintenance than conventional electricity and gas meters and consumers are not usually metered individually but collectively in groups of dwellings. Paying a fixed share of an aggregate charge tends to be unpopular with consumers and does not encourage individual economy in use. On the whole, the costs to consumers within CHP schemes are not significantly lower than the best conventional systems for heating individual dwellings. The benefits in terms of primary fuel saving are considerable but these accrue to the nation as a whole.

The Marshall Report (DEn, 1979) and the Atkins Report (DEn, 1984) recommended some commitment to CHP in the UK, initiated by pilot schemes in inner-city areas (including Belfast, Glasgow, Leicester, Newcastle upon Tyne, Sheffield and the London Boroughs of Southwark and Tower Hamlets) which could be evaluated as

they were implemented over a period of 20 to 25 years. However, in addition to the points made above, the ready availability of fossil fuels, the increasing costs of capital investments and a preference by central government for extending consumer choice have all militated against large-scale CHP developments in the UK.

In many parts of northern Europe, especially where the winter heating season is longer and more intense than in the UK, district heating schemes with CHP generation are the norm rather than the exception. In Sweden, for example, high energy efficiency is achieved by using the district heating network in a manner analagous to an electricity distribution grid. In one such scheme at Malmo, about half of the city's heating requirements are supplied by a small CHP station which generates electrical power and heat in an average ratio of one to two from imported oil or gas at an efficiency of 86%. While the output from the CHP station broadly follows the varying seasonal demand, the base load year-round requirement for hot water is drawn from a variety of smaller sources of which the largest is the municipal refuse incinerator. Here the higher costs (including stringent control of gaseous emissions) compared with disposal by landfill are offset by the value of the heat recovered. Other continuously available sources include the local sewage disposal plant (making use of heat pumps to recover energy) and many industrial and commercial undertakings that have combustible waste products (including the waste straw and manure from the stables of the local racecourse). In the coldest weather, a coal-fired heat-only station operating at high efficiency (80–90%) and having a maximum output about one-fifth as great as the CHP station can be called upon to top up the system. The extensive network of district heating pipes means not only that new consumers can be plumbed in as the city expands but that new industrial and commercial suppliers can also be connected, which is to their own financial advantage. At any one time, the operating municipal authority can select the cheapest overall mix of energy sources available with obvious benefit to its customers. As already mentioned, it is the local ownership and control of energy utilities that greatly facilitates this type of operation.

As mentioned above (chapter 6), there are a small number of installations in the UK (at Nottingham, for example) which supply district heating and

Table 9.1   **The matching of energy applications with the most appropriate sources**

Source	Energy quality	Examples of application
Heat > 2000°C Nuclear power Storm winds and waves Concentrated sunlight	Very high	Industrial furnaces
Heat > 1000–2000°C Fossil fuels Food Bright sunlight	High	Motive power
Heat 100–1000°C Natural geysers Fresh breeze Fast-flowing river	Moderate	Electricity generation Cooking
Heat < 100°C Hot aquifers	Low	Space heating

Source: Miller (1990)

electricity from the incineration of municipal refuse. Some industrial firms with a demand both for electrical power and process heat operate their own small CHP stations, a development often known as *co-generation*. They are designed generally to follow the demand for heat, with the surplus or deficit of electricity sold to or purchased from the national grid. This type of arrangement with a national utility was made possible by the 1983 Energy Act. As much as 5% of the UK's electricity supply is now generated by industrial producers (about a quarter of which is nuclear power from the UK Atomic Energy Authority and British Nuclear Fuels plc) and a third of this is delivered to the public supply.

A similar two-way arrangement with the national grid at the even smaller scale of mini-CHP (minichip) allows institutions such as hotels and hospitals with a high and continuous demand for heat and electricity (including, for example, the Devon and Somerset Police Headquarters at Exeter) to generate their own heat and power at an overall efficiency of about 80%. Small units, based on a car engine modified to burn natural gas, have electrical power ratings of 5 kW or greater, with an output of heat up to three times as large.

In the longer term: essential electricity

A more radical approach to the supply and use of energy involves paying particular attention to the closer matching of energy requirements with the

most appropriate source (table 9.1). For example, electricity is an extremely convenient and widely available source of energy in MDCs and is therefore employed in a wide variety of applications; but, as has been mentioned many times, its generation from fossil fuels in conventional power stations is an inefficient process. A popular application, on account of its low capital costs, flexibility and cleanliness, is for domestic space heating; but here, after making further allowance for transmission and other losses, only about a quarter of the energy content of the primary fuel appears as useful heat for the consumer. To put it another way, a high-grade energy source (e.g. coal or oil) is being used to satisfy a low-grade need (space heating). Ideally, such a demand should be matched to a low-grade source, such as passive or active solar heating or, as in CHP applications discussed above, the reject heat from power stations. More realistically perhaps, the overall efficiency could be raised to about 75% by using the coal or oil directly in a modern type of domestic central heating boiler.

In the UK, 15% of delivered energy is in the form of electricity of which about two-thirds may be considered essential uses: these include lighting, domestic and other appliances, industrial motive power and electrochemical processes. It has been suggested that the reduction or elimination of non-essential mainly thermal uses of electricity would remove the need to increase national generation capacity and decrease the urgency for replacement of existing power stations that are approaching the end of their useful lives. Likewise, more time would be available for the development of renewable sources of electrical power. Together with improved efficiencies in electricity consumption in essential uses, as referred to above, significant savings could result.

Some commentators would stress the desirability of combining this approach with further measures, such as a reduction in the consumption of raw materials in our 'throw-away' society and a reorganization of patterns of work to decrease unnecessary travelling by commuters. It is not clear how such a transition towards a 'sustainable-earth' society, which questions the whole basis of modern industrial economies, would come about and it must be seen, even by its advocates, as a long-term prospect.

Perhaps even more important than the use of electricity in MDCs is its rôle in the LDCs. Attention has already been drawn (chapter 8) to the environmental impact of large hydropower schemes (in Amazonia, for example) and many observers have questioned the wisdom of such 'prestige' investments, bearing in mind their cost and the socio-economic implications of the uses to which such large quantities of power might be put. Some LDCs (Zimbabwe, for example) regard the widespread provision of an electricity supply from a national grid as an important political goal, one which perhaps pre-empts possible openings for the utilization of sources of renewable energy and may contribute to irreversible cultural change.

Points to consider

- What scope is there for conservation of energy in (a) your home (b) your school or college (c) your personal lifestyle?

- Consider your energy-consuming activities and requirements and identify which of them can be provided only by electricity.

- What obstacles exist to the expansion of the reclamation and recycling of materials?

- By what means might fuel consumption by private motorists and road hauliers be reduced and would such action be justified?

- At what scales, if any, is CHP technology suited to extensive deployment within the UK?

- Is energy conservation either necessary or desirable for MDCs?

- Does energy conservation have any relevance for LDCs?

- Does the future well-being of the planet depend upon fundamental changes in attitudes to the exploitation of energy and other resources? If not, why not? If so, do they apply equally to MDCs and LDCs? How might such changes be brought about?

10 Resources for a changing world

- Tomorrow, or a year hence, we may propose and test important theories of which nobody has seriously thought so far. If there is growth of knowledge in this sense, then it cannot be predictable by scientific means. For he (*sic*) who could so predict our discoveries of tomorrow could make them today.

(Karl Popper, 1972)

In so far as the exploitation of the Earth and its resources are concerned, people may be said to tend towards one of two contrasting points of view, which may be conveniently labelled *cornucopian* or *catastrophist* (Cotgrove, 1982).

The *cornucopian viewpoint* considers that, as in the past, the Earth's resource base will continue to expand and will always be sufficient for people's needs. Scientific and technological advances will continue to provide, and indeed constitute a moral imperative to secure, economic prosperity and an increasing standard of living in the MDCs and the eventual spread of such benefits to the LDCs. The magnitude of any difficulties related to the decreasing availability of resources, or increasing environmental impacts of their exploitation, tends to be exaggerated and such problems are in any case local or, if large-scale, temporary, since they

can always be overcome. The principal energy resources will continue to be the fossil fuels, with unconventional sources of oil developed if necessary, and nuclear power, including breeder (fission) and probably fusion reactors. The cornucopian point of view emphasizes the progress achieved, especially during the last century or so, in improving both the physical and material well-being of an increasing number of people and would regard this as a triumph for humankind over the untamed forces of nature. The future promises continuing economic growth and unlimited improvements in the quality of human existence. Attitudes towards the natural world incline towards domination, control and exploitation. Opponents of this viewpoint see it as unrealistically optimistic and objectionably technocratic; it has been labelled 'business as usual' and 'technological fix'.

The *catastrophist point of view*, on the other hand, follows Malthusian thinking in warning of an inexorable depletion of finite resources by an increasing population, accompanied by inevitable environmental degradation. The rates of resource use and creation of waste occurring in MDCs are seen as unjustifiable, socially divisive, immoral and ultimately unsustainable. Technological solutions to the increasing number and variety of environmental problems are at best temporary and often counter-productive in that they may create more problems than they solve and, in the face of economic and institutional obstacles, they frequently fail to be implemented. Renewable energy resources should be rapidly developed and widely deployed; fossil fuels should be conserved as an invaluable feedstock; nuclear power should be abandoned as both uneconomic and unacceptably hazardous. A catastrophist perspective points to a growing disparity between rich and poor both within and between nations and the extensive damage and loss caused to living and non-living resources by unrelenting economic growth. Future survival depends on conservation by the affluent and sustainable development for the impoverished. Attitudes to the natural world stress harmony, co-operation and guardianship. Opponents of this point of view regard it as exaggeratedly pessimistic and dispiritingly gloomy; it has been branded 'the doomsday syndrome' and 'spaceship earth'.

This brief synopsis of contrasting ideas on the proper relationship between people and their environment serves to underline that the important issue of the securing of a dependable and sufficient energy supply for the foreseeable future is part of that wider concern. It is not part of the purpose of this volume to attempt a judgement between, or reconciliation of, the two opposing perspectives outlined above. But when considering the future pattern of energy resource development in this and other possibly very different countries, you will surely be aware of your own views and values and of the extent to which attitudes are likely to be affected by the personal circumstances of the observer. The environment will look very different from Burkina Faso and Birkenhead, Djibouti and Gibraltar, Mali and Marlow, San Francisco and San Salvador.

It must also have become clear that any problems and issues concerned with the development and use of energy resources cannot be treated in isolation because of their interrelations with other activities impinging upon the environment and with wider demographic, economic, institutional, political and social considerations; everything is connected to everything else! The future pattern of energy sources and demands is just a part of much wider issues concerning marked global inequalities in living standards and significant changes to the natural environment. It is quite unrealistic to suppose that disagreements over the future direction of energy policy could be solved by technical argument, but even the unavoidable value judgements ought to be informed as far as possible by a basis of fact and understanding.

It is also necessary to remember that, even if agreement were to be achieved on a future energy policy for any particular country, wide differences of opinion would undoubtedly arise regarding the means by which such goals could or should be attained.

In conclusion, it remains only to stress that, in the field of energy resource studies – as in any other – new and as yet unimagined ideas and approaches will appear; many, surely, from amongst the rising generation. It is important, then, to be able to survey them from a secure and informed foundation, but with a receptive mind able to respond and, if necessary, modify firmly held beliefs. After all, the only predictable aspect of the future is that it will surely be different from what we expected!

Points to consider

• Draw up a large table comparing the relative merits of the different types of energy resource discussed in this book. Some of the headings you might use include: nature of primary energy source, present and future availability (quantities and locations), energy density, range and efficiency of possible conversions and applications (including any outside the area of energy utilization), firmness of supply, state and possible scales of technological development, appropriateness to level of economic development, relative magnitudes of capital investment and operating (fuel) costs, present and likely future costs, range and degree of environmental impacts in both exploitation and utilization (to land, water, air, plants, animals, people, structures, amenity), commercial attractiveness, public acceptability, institutional and governmental appeal.

- Which of the many headings that you used in the preceding exercise do you consider to be the most important in practice?

- What, if anything, should be the function of a national energy policy?

- Assuming that, as recent experience suggests, fossil fuels will be available in sufficient quantities into the foreseeable future, what relevance have either nuclear power or renewable energy resources for the more developed world at the present time?

- Comment on the view that the real energy crisis concerns fuelwood and the LDCs, not fossil fuels and the MDCs.

- Does the maxim 'think globally, act locally' have relevance or usefulness in the context of energy resource studies?

- How true is it that 'energy is the issue that today drives nations apart and causes wars'? (*Radio Times*: Radio 4, 30 March 1991)

- Comment on the assertion that 'the problems of the next century will not be pollution, but the quantity of energy we handle. Yet no one in the western world wants to question industrial growth.' (Keith Scholey in *Radio Times*: BBC 2, 2 February 1992)

- Read again the quotation at the beginning of the first chapter. Do you agree that 'future development crucially depends upon [energy resources] that are dependable, safe, and environmentally sound'? Do you consider that 'at present no single source or mix of sources is at hand to meet this future need'? In the light of your answer, what action should be taken?

As a final exercise you are invited to analyse and compare the availability and use of energy resources in different countries. Although there is space only for limited data in this book, you could extend the scope of your work using the same or supplementary sources of information. For the MDCs data are reasonably detailed and reliable: a sample is given in appendix 4. Some of the topics you could investigate include: the relative proportions of the different energy resources used; the extent of reliance on imported energy; the proportions of energy consumed by different sectors of the economy; the efficiency of energy use in relation to both economic activity and population size; the allocation of expenditure on research, development and demonstration (R, D & D) between different energy sources; the amount of R, D & D expenditure relative to both economic activity and population size. You will find it helpful to compute appropriate *key indicator ratios*, for example total primary energy requirement (TPER) per capita, oil use / GDP, total final consumption (TFC) / GDP or R, D & D per capita. As part of your investigations, you could try to explain past trends and justify future projections, and account for any differences between (otherwise similar) countries.

Data for LDCs are generally less detailed and more imprecise: a selection is presented in appendix 5. In attempting to interpret these data it is important to take note of energy resources not specifically listed: references may be made to material from earlier chapters and to supplementary climatological, oceanographic, geophysical or other data as appropriate.

Appendix 1: resources and reserves

It needs to be stressed that material objects are not intrinsically resources; they are such only in so far as they perform a useful function or have value to people.

> Natural resources are products of the human mind; their limits are not physical, but are set by human demands, institutions, imagination and ingenuity (Rees, 1985, p.396).

So what is or is not regarded as a resource will change with the passage of time and will be different for different societies and cultures. To take a simple example, uranium ore was not considered to be a resource anywhere a century ago and neither is it today for native peoples in the remote Amazonian rain forest.

In the particular case of mineral resources, the *resource* may be regarded as the totality of a substance present, or believed to be present, on the Earth, while the perhaps more useful concept of a *reserve* is defined as that fraction of the resource whose location is known or inferred and which is available for exploitation with present technology and under current economic conditions.

As shown in fig. A1.1, it is helpful to qualify the size of reserves available with adjectives such as proven or probable, reflecting the degree of geological certainty concerning their existence. Clearly, further investigation and exploration will result in the movement of boundaries between

categories along the horizontal axis. Likewise, boundaries will move in a vertical direction depending on advances in the technology of extraction and utilization or on fluctuations in demand; either of these could render previously marginal reserves recoverable, or consign currently recoverable reserves to the marginal category.

Finally, it should be pointed out that there is a lack of agreed and consistent usage of terminology between, and even sometimes within, different branches of the mineral industry. Thus it is necessary to proceed with some caution in the evaluation of statements on resource availability. In the petroleum industry, for example, the use of the terms *proven, possible* and *probable* relates to the estimated statistical probability of the reserves being economically and technically producible. The probability levels are respectively > 90%, > 50% (but ≤ 90%) and > 10% (but ≤ 50%).

It also needs to be borne in mind that, in addition to natural uncertainties, figures can be distorted by extra-technological business considerations: for example, there is a general tendency for the proven reserves of petroleum declared at the time of discovery to be revised upwards as exploitation proceeds. This is known as the phenomenon of *reserve appreciation.*

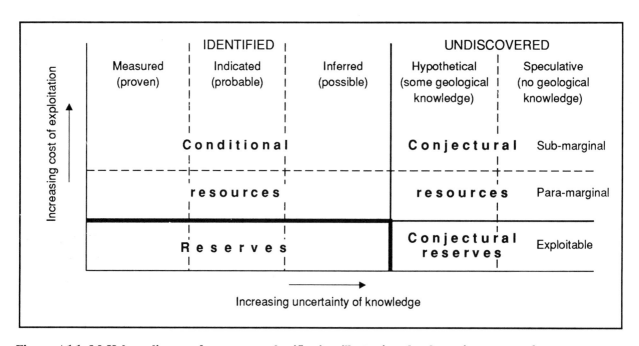

Figure A1.1 McKelvey diagram for resource classification illustrating the dynamic concept of a resource.

Appendix 2: classification of resources

Resources of all types, including energy, are conventionally separated into *non-renewable* and *renewable* types, or *stock* and *flow* resources. Strictly speaking, all resources are renewable since they are ultimately the product of natural cycles, and matter can neither be created nor destroyed. The distinction between renewable and non-renewable lies in the time period over which replenishment takes place. The range of time scales is indeed immense: perhaps many billions of years for uranium, tens of millions of years for fossil hydrocarbon fuels, thousands of years for landscapes, centuries for peat and soils, a few seasons for vegetation, months for hydropower, hours for tidal power, seconds for wind power and the merest fraction of a second for sunlight. It is the human timescale, then, which determines the degree of renewability of a resource.

A basic separation is shown in fig. A2.1 where both stock and flow types have been further subdivided. Within stock resources, it is helpful to make a distinction between those, like fossil fuels, that are totally consumed by use and others, such as metallic minerals, that are not irrevocably transformed by use and can in principle be recovered and recycled without limit. This may of course be economically or technically impractical, and is likely to be so for many non-metallic minerals, for example china clay and phosphate.

Similarly, flow resources may be classified according to the extent that human activities can influence their availability. At one extreme, a resource such as sunlight reaches the Earth at a rate totally beyond the control of people; such resources are often designated *continuous*. Other flow resources, such as fertile soils and living species of plants and animals, may be seriously degraded or depleted by human activities, even to the point of non-renewability.

The most familiar types of energy resource, fossil fuels and uranium, are entirely consumed by use and therefore non-renewable; most of the others are renewable subject to varying degrees of human influence. A possible classification of energy resources is depicted in fig. A2.2, where a further subdivision has been made on the basis of their direct or indirect dependence, or otherwise, on the Sun.

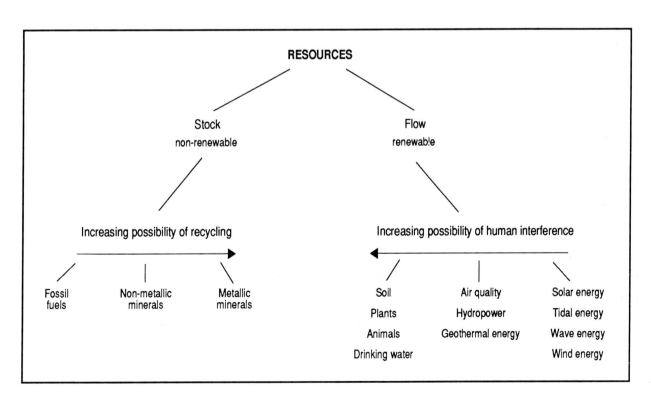

Figure A2.1 Types of resource classified according to their renewability.

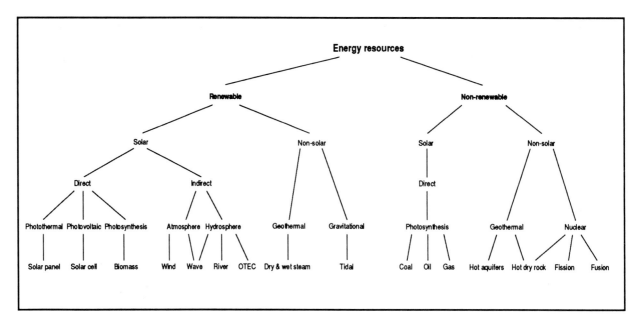

Figure A2.2 A possible classification of energy resources.

Appendix 3: nuclear radiation and people

Although SI units relating to radiation have been in use for a number of years, references to the system they replaced are still common, hence both will be included in the brief definitions which follow.

The *becquerel* (Bq) is the SI unit of activity, formerly the *curie* (Ci). A transformation rate of one per second (through α- or β-decay) constitutes an activity of 1 Bq. The curie was defined as the number of transformations undergone in one second by one gram of natural radium; hence 1 Ci = 3.7 x 10^{10} Bq.

The *gray* (Gy) is the unit of absorbed dose and is defined in terms of the amount of energy imparted to the absorbing medium. A dose of 1 Gy is equal to the deposition of 1 J of energy by ionizing radiation within 1 kg of material (e.g. human tissue). The former unit was the *rad* (from *r*adiation *a*bsorbed *d*ose) and 1 rad = 100 Gy.

Radiation doses were formerly expressed in terms of *exposure*, the quantity of radiation directed towards a target, which is not necessarily the same as the amount absorbed. The standard unit is the *roentgen*, defined as the quantity of electro-magnetic radiation (X- or γ-rays) that produces 258 μC of electric charge, of either sign, per kg of air at standard temperature and pressure (originally 1 electrostatic unit or e.s.u. per cm^3). Because the roentgen is strictly applicable only to X- or γ-radiation below 3 MeV in air, and because it measures exposure rather than absorbed dose, the rad was recommended for universal use in 1956 (although the roentgen is still generally current in the former USSR, for example). As far as human tissue is concerned, the roentgen and the rad are equivalent to within about 5%.

The *sievert* (Sv) is the SI unit of dose equivalent. The need for such a measure arises since equal absorbed doses from different types of radiation do not necessarily have equal biological effects. Hence a dose equivalent of 1 Sv is equal to an absorbed dose of 1 Gy weighted according to its biological effectiveness. The weighting factor is 20 for α-particles, 10 for neutrons and unity for β-particles, γ-rays and X-rays. The former unit of dose equivalent was the rem (from *r*oentgen *e*quivalent *m*an) and 1 Sv = 100 rem. Further weighting factors are employed to allow for the enhanced susceptibility of particular organs to particular types of damage: e.g. bone marrow (cancer) and gonads (genetic). The result of such weighting yields an *effective* dose equivalent.

A significant and perhaps unfortunate feature of low doses of radiation is that the recipient is quite unaware of them and they result in no immediately detectable effects in the human body (fig. A3.1). Acute injury requires a dose in excess of some threshold level, about 500 mSv for the average person; an instantaneous dose equivalent of about 5000 mSv would most probably be fatal within a matter of days. Because of cellular repair mechanisms, the human body is able to tolerate large radiation doses when spread over a sufficiently long period of time. There is, however, no threshold for the induction of long-term somatic (e.g. cancer) or genetic (hereditary) effects. In this sense, no dose of radiation, the natural background not excepted, is 'safe'. The mechanisms that result in damage are not fully understood in detail, however, and the long time-scale militates against establishing specific cause-and-effect relationships (fig. A3.1). It is agreed, however, that higher doses of radiation increase the risk of damage becoming manifest at some time in the future.

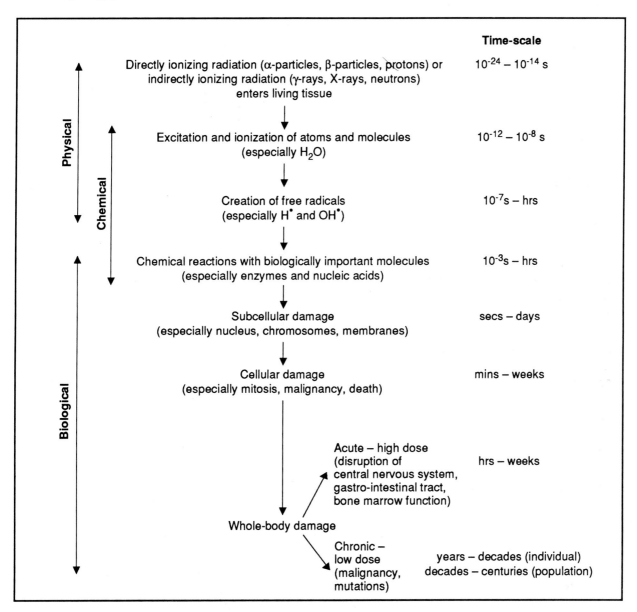

Figure A3.1 The chain of events leading to radiation damage in people.
Source: adapted from Coggle (1983)

Appendix 4: energy data for selected OECD

	Belgium				Netherlands				Sweden			
	1973	1979	1987	2000	1973	1979	1987	2000	1973	1979	1987	2000
Energy production (Mtoe)												
Coal	6.4	4.4	3.1	1.9	1.3	–	–	–	–	–	–	–
Oil	–	–	–	–	1.6	1.6	4.8	4.2	–	–	–	–
Gas	–	–	–	–	54.8	72.1	57.2	47.5	–	–	–	–
Nuclear	–	2.5	9.4	9.8	0.2	0.8	0.8	4.5	0.5	4.7	15.1	13.3
Hydro/geothermal	0.1	0.1	0.3	0.1	–	–	–	–	13.4	13.7	16.3	14.8
Other renewable	–	0.1	0.2	–	–	0.2	0.1	1.4	3.5	3.9	5.6	5.8
Net imports (Mtoe)												
Coal	4.6	6.2	5.2	9.4	1.7	3.8	7.0	24.9	1.7	1.8	2.5	3.9
Oil	28.3	27.0	19.4	18.0	29.8	31.8	17.1	19.5	27.8	28.6	15.0	19.2
Gas	7.3	9.5	7.6	9.3	-25.7	-38.7	-22.9	-19.0	–	–	0.2	1.0
Electricity	-0.1	-0.1	-0.2	-0.2	-0.1	–	0.3	n.a.	0.1	0.1	-0.3	0.0
Total primary energy requirement (TPER)	46.7	48.4	45.4	48.3	63.3	69.9	65.9	73.0	47.4	51.6	56.4	58.1
Energy consumption (Mtoe)												
Industry	19.8	17.7	13.9	13.7	22.4	25.2	22.7	27.4	15.9	15.1	13.8	15.8
Transport	5.0	6.0	7.0	8.1	7.5	8.6	9.5	8.9	5.5	6.3	7.2	7.9
Other (mainly buildings)	12.9	14.6	12.6	12.3	20.4	22.8	21.3	19.6	14.3	15.3	13.6	13.7
Total final consumption (TFC)	37.7	38.3	33.5	34.1	50.3	56.6	53.5	55.9	35.7	36.7	34.6	37.4
GDP (billions of $US)	64.7	73.8	82.8	120.4	107.4	118.8	130.3	194.5	81.0	90.0	103.6	137.8
Population (millions)	9.7	9.8	9.8	9.9	13.4	14.0	14.7	15.6	8.1	8.3	8.4	8.5

Spending on energy R D & D (millions of $US)

	Belgium		Netherlands		Sweden	
	1979	1988	1979	1988	1979	1988
Fossil HC	6.5	3.5	12.0	32.6	16.2	7.3
Nuclear fission	90.0	56.7	93.2	20.9	19.1	4.9
Nuclear fusion	3.3	12.2	16.8	10.8	7.2	10.0
Renewable						
Photothermal		*1.1*		*4.2*		*6.0*
Photoelectric		*0.8*		*3.0*		*0.2*
Biomass		*0.2*		*5.0*		*9.0*
Ocean		*–*		*1.0*		*–*
Wind		*0.2*		*6.4*		*2.8*
Geothermal		*0.8*		*2.1*		*0.2*
Total renewable	6.4	3.1	20.9	21.7	66.4	18.1
Conservation	8.4	3.0	26.4	45.7	45.3	23.7
Other	8.7	2.3	20.5	30.5	24.0	22.0
Total	123.3	80.8	189.8	162.2	178.2	86.0

Source: IEA (1989c)

countries

	Australia				Japan				UK		
1973	*1979*	*1987*	*2000*	*1973*	*1979*	*1987*	*2000*	*1973*	*1979*	*1987*	*2000*
40.3	50.8	98.8	124.5	17.9	10.6	7.5	5.5	75.9	69.8	60.2	n.a.
19.8	22.7	28.2	13.3	0.8	0.6	0.7	2.5	0.6	79.8	126.7	
3.4	7.1	12.9	24.1	2.3	2.2	1.9	4.6	24.9	33.6	40.1	
–	–	–	–	2.2	15.7	41.9	79.2	6.3	8.6	12.3	
2.6	3.6	3.3	4.0	16.1	19.2	18.4	28.7	1.0	1.2	1.4	
3.5	3.5	4.0	4.8	–	–	–	1.7	0.2	–	–	
-17.6	*-25.1*	*-62.0*	*-79.2*	40.9	39.8	59.8	84.2	*-0.9*	1.0	5.2	
7.4	9.5	1.4	22.7	254.8	263.1	210.9	203.3	110.6	16.6	*-51.4*	
–	–	–	-7.0	2.8	17.0	34.8	44.8	0.7	7.6	10.2	
–	–	–	–	–	–	–	–	–	–	1.0	
57.3	71.1	80.8	107.2	331.3	362.7	371.7	454.5	221.1	220.1	208.7	
17.6	20.0	21.4	25.3	159.6	147.5	130.5	131.8	70.9	61.1	46.2	
13.5	17.2	20.5	25.6	40.9	53.5	62.0	75.9	31.0	33.8	40.8	
9.2	9.5	11.1	15.7	51.7	60.9	70.4	94.6	51.4	57.9	59.3	
40.3	46.7	53.0	66.6	252.2	261.9	262.9	302.3	153.3	152.8	146.3	
113.6	132.9	167.3	238.2	847.7	1047.9	1416.2	2287.7	385.5	421.0	484.1	
13.5	14.5	16.3	19.1	108.7	115.9	122.1	131.0	56.2	56.2	56.9	

	Australia		Japan		UK	
	1979	*1988*	*1979*	*1988*	*1979*	*1988*
	22.3	28.5	89.1	387.1	84.4	28.3
	21.5	15.8	1343.9	1245.2	358.8	229.3
	–	2.1	213.4	240.5	43.9	43.9
				4.4		*3.4*
				54.2		*–*
				23.7		*3.4*
				0.8		*2.1*
				2.9		*7.3*
				42.5		*5.3*
	10.7	0.9	71.5	128.5	32.4	21.5
	5.5	4.6	49.8	63.5	32.9	36.8
	7.2	1.3	42.7	94.1	53.3	38.4
	67.2	53.2	1810.4	2158.9	605.7	398.2

Appendix 5: energy data for selected non-OECD countries

	1	2	3	4	5	6	7	8	9	10	11	12	13	14	15	16
Africa																
Ethiopia	49.5	120	10	20	0.01	0.03	–	–	0.08	0.12	–	–	650	0.8	27	27.2
Gabon	1.1	2960	153	1155	–	–	–	–	0.017	–	19	29	33	0.4	21	1.9
Mali	8.2	270	10	24	–	0.003	130	–	–	–	–	–	–	0.05	7.2	3.5
Morocco	24.5	880	124	244	0.05	0.18	0.3	–	0.002	0.005	–	–	4.5	0.7	3.6	0.9
Mozambique	15.3	80	81	84	0.24	1.90	–	–	0.065	–	–	–	50	3.2	16	10.2
Nigeria	113.8	250	34	135	0.2	1.4	2200	n.a.	2.4	4.6	–	–	2.2	1.9	15	67.1
South Africa	35.0	2470	1744	2432	55	126	20.1	–	0.03	–	430	2100	–	0.5	1.3	5.1
Zambia	7.8	390	464	372	0.06	0.09	–	–	–	–	–	–	34	2.0	30	6.7
Latin America																
Argentina	31.9	2160	975	1718	0.1	8	300	5400	0.7	0.9	11.8	385	200	11.0	45	4.1
Cólombia	32.3	1200	413	754	9.7	17	200	n.a.	0.1	–	–	–	1290	7.7	52	10.8
Costa Rica	2.7	1780	267	614	–	0.05	–	–	–	–	–	–	223	0.8	1.8	1.9
Jamaica	2.4	1260	703	902	–	–	–	–	–	–	–	–	0.5	0.03	–	0.01
Nicaragua	3.7	790	172	259	–	–	–	–	–	–	–	–	17	0.1	4.5	2.1
Panama	2.4	1760	576	1636	–	–	–	–	–	–	–	–	16	0.6	4.2	1.2
Paraguay	4.2	1030	84	226	–	–	–	–	–	–	–	–	25	8.5	20	3.6
Uruguay	3.1	2620	765	779	–	–	–	–	–	–	–	–	>10	1.0	0.6	2.2
S.E. Asia																
Burma	38.0	200	39	76	0.002	0.2	8	n.a.	0.3	–	–	–	366	0.3	32	11.8
Malaysia	17.4	2160	312	920	0.004	0.7	400	n.a.	1.5	2.4	–	–	123	1.5	21	5.6
Nepal	18.4	180	6	24	n.a.	4	–	–	–	–	–	–	364	0.3	2.1	11.4
Sri Lanka	16.8	430	107	173	17	22	–	–	–	–	–	–	10	0.9	1.8	5.8
E. Europe																
Hungary	10.6	2590	1825	3106	4.5	16	40	400	0.1	0.4	10	50	7.5	0.5	1.6	2.1
Poland	37.9	1790	2027	3333	40	200	2	5	0.1	0.2	10	50	23	0.8	8.6	2.7
Romania	23.2	–	1536	3514	n.a.	4	200	n.a.	0.2	–	50	100	70	4.6	6.2	3.3
Yugoslavia	23.7	2920	898	2241	17	22	30	n.a.	0.08	–	–	–	116	8.0	9.1	3.2

1 Population in millions (1989) **2** GNP per capita $ (1989) **3** Energy consumption per capita in kgoe (1965) **4** Energy consumption per capita in kgoe (1989)
5 Coal: proven recoverable reserves in Gt **6** Coal: estimated amount in place in Gt (includes **5**) **7** Oil: proven recoverable reserves in Mt
8 Oil: estimated amount in place in Mt (includes **7**) **9** Gas: proven recoverable reserves in Tm^3 **10** Gas: estimated amount in place in Tm^3 (includes **9**)
11 Uranium: proven reserves in kt ($< \$130\ kg^{-1}$) **12** Uranium: estimated additional reserves in kt (includes **11**) **13** Hydropower: gross theoretical capability TWh a^{-1}
14 Hydropower: installed capacity (including plant under construction) in GW **15** Biomass: total forest area in Mha **16** Biomass: fuelwood (including charcoal) production in Mt a^{-1}

Source: World Bank (1991); WEC (1989)

Glossary

actinides   The name for the group of elements with atomic numbers from 89 to 103; all are radioactive. The first four (actinium, thorium, protactinium, uranium) occur in nature; the remaining transuranic elements are formed by nuclear reactions, including those occurring in nuclear reactors.

alpha-decay   The spontaneous emission of an alpha particle (*q.v.*) by a heavy radionuclide (*q.v.*), e.g. $_{92}U^{238} \rightarrow {}_{90}Th^{234} + {}_2He^4$ (α).

alpha-particle (α)   The nucleus of the helium-4 atom (symbol $_2He^4$), composed of two protons and two neutrons.

alpha-radiation   A stream of alpha-particles (*q.v.*), emitted by certain radionuclides (*q.v.*). Typically they are stopped by 40 μm of tissue or a few cm of air. Since they are unable to penetrate ordinary clothing they present little hazard to humans unless inhaled, ingested or injected into the body.

beta-decay   The spontaneous emission of a beta-particle (*q.v.*) by certain radionuclides (*q.v.*), e.g. $_6C^{14} \rightarrow {}_7N^{14} + {}_{-1}e^0$. The beta-particle (electron) originates from the nucleus of the atom and involves the transformation of a neutron into a proton, thus $_0n^1 \rightarrow {}_1p^1$ ($_1H^1$) $+ {}_{-1}e^0$ (β).

beta-particle (β)   An electron, symbol $_{-1}e^0$. (The term also includes positrons, $_{+1}e^0$.)

beta-radiation   A stream of energetic beta-particles, emitted by certain radionuclides (*q.v.*). Typically they are stopped by 4 cm of human tissue or a few mm of aluminium or plastic.

char   The solid residue arising from the pyrolysis (*q.v.*) of organic materials. Its high calorific value makes it useful as a solid fuel.

cullet   Term used in the glass industry to denote waste fragments, offcuts, mis-firings, substandard products, etc., which are colour sorted and suitable for remelting. Most cullet arises within the industry itself; the cost of retrieval from the domestic and commercial waste stream (e.g. bottle banks) may exceed the product value.

energy   the capacity (of materials or radiation) to perform work. The standard SI unit of measurement is the joule *(q.v.)*.

free radical   A molecule (ion or atom) having one (or more) unpaired electrons, which renders it highly chemically reactive. It is denoted by a dot superscripted to the chemical symbol(s), e.g. H˙, OH˙.

Fresnel lens   A lens composed of many narrow stepped concentric rings having the advantage of being thinner and less massive than a normal convex lens of equivalent focal length. They are commonly used in spotlights, viewfinders and on the rear windows of public service vehicles.

gamma-radiation (γ)   Electromagnetic radiation of extremely short wavelengths (0.1–100 pm). Propagating as waves, in its interactions with matter it is better regarded as a stream of high energy (0.1–100 MeV) photons (*q.v.*). It is progressively attenuated by heavy nuclei; typically, 4 cm of lead reduces the incident intensity to one tenth.

isotope   Nuclides (*q.v.*) having the same number of protons (i.e. same atomic number) but differing numbers of neutrons (hence differing mass numbers) in their nuclei. Isotopes of a given element have identical chemical properties but different physical properties. Hence the enrichment of natural uranium in the uranium-235 isotope can only be performed by (difficult) physical processes.

joule (J)   the SI unit of work or energy (q.v.) It is defined as the amount of energy required to move a force of one newton through a distance of one metre in the direction of the force.

nucleus   (1) The dense central positively charged part (diameter $\sim 10^{-14}$ m) of an atom (diameter $\sim 10^{-10}$ m). It is composed of protons and neutrons, and is surrounded by a cloud of atomic electrons. (2) An ovoid membrane-bound structure (organelle) in living cells containing the chromosomes and controlling cellular activity.

nuclide   A species of atom characterized by the number of protons (the atomic number) and the total number of protons and neutrons (the mass number) in the nucleus. Nuclides are specified by a chemical symbol corresponding to the atomic number, which is appended as a subscript, together with the mass number as superscript. Thus $_{92}U^{238}$ (uranium-238) is the isotope (q.v.) of the ninety-second element in the Periodic Table whose nucleus has 92 protons and 146 (= 238 − 92) neutrons. Since atoms are electrically neutral, there must also be 92 extra-nuclear electrons.

osmosis The passage of a solvent through a semi-permeable membrane from a more dilute to a more concentrated solution, a process which tends to equalize the concentrations.

photon A quantum (discrete packet) of electromagnetic radiation. It carries no charge and has zero mass (except by virtue of its energy content). The energy of a photon is given by $E = hf = hc/\lambda$ where f, c and λ are respectively the frequency, velocity (normally 3×10^8 m s^{-1}) and wavelength of the radiation, and h is Planck's constant (6.626×10^{-34} J s).

power the rate of using or supplying energy (q.v.). The standard SI unit of measurement is the watt (q.v.).

pyrolysis The decomposition of a substance by heat in the absence of oxygen.

radionuclide An atom with an unstable nucleus which spontaneously disintegrates with the emission of an alpha- or beta-particle and/or a gamma-ray.

recycling The reuse of waste materials as inputs in the economy. Though energy itself cannot be recycled, the recycling of waste materials can reduce overall consumption of energy in the economy and may, by reducing the volume of waste requiring ultimate disposal, be desirable on environmental grounds. A useful distinction (in roughly decreasing order of economic efficiency) may be made between:

(a)   reuse with no product modification e.g. returnable milk bottle

(b)   direct recycling with reuse in the same form and with the possibility of further recycling e.g. glass cullet (q.v.), scrap paper and plastics;

(c)   indirect recycling with reformulation into a new product, usually with no further possibility of recycling, e.g. pulpboard made from paper/plastics/wood;

(d)   recovery of energy/nutrients in which the waste product is subject to irreversible processes (anaerobic digestion, composting, fermentation, incineration, pyrolysis (q.v.) etc.) with the release of useful energy/nutrients.

semiconductor A substance, such as pure crystalline silicon, having an electrical resistance intermediate between those of conductors and insulators is an *intrinsic* semiconductor. Electrical conductivity is markedly affected by the absorption of energy (heat or light) and the 'doping' of silicon with suitable impurities to give *extrinsic* semiconductors enhances this phenomenon. The presence of a small number of phosphorus atoms in the silicon lattice will create an excess of electrons (n-type) while doping with boron results in an excess of positive 'holes' or sites lacking electrons (p-type). An n-p type photovoltaic cell has a thin (~ 0.5 μm) layer of n-type material overlying p-type material (~ 300 μm thick). Light (photons) absorbed at the junction between the two types of semiconductor materials creates electron-hole pairs which then move in opposite directions, so giving rise to an electric current in an external circuit.

Sv (sievert) The SI unit of dose equivalent for ionizing radiation (see appendix 3).

watt (W) the SI unit of power (q.v.) It is defined as a rate of working or of energy use or delivery equal to one joule (q.v.) per second.

X-radiation Electromagnetic radiation of very short wavelengths (0.01–10 nm). Propagating as waves, in its interactions with matter it is better regarded as a stream of energetic (0.1–100 keV) photons (q.v.).

Bibliography

Addinall E. and H. Ellington (1982) **Nuclear Power in Perspective** Kogan Page, London

Agius P. J. V. (ed.) (1979) **A Warmer House at Lower Cost** The Watt Committee on Energy, London

Ashby M. (1979) **Water Power in the Dyfi Valley** Centre for Alternative Technology, Machynlleth

Battarbee R. *et al.* (1988) **Lake Acidification in the UK 1800–1986** ENSIS, London

Blowers A. and D. Pepper (eds.) (1987) **Nuclear Power in Crisis** Croom Helm, London

Bockris J. O'M. (1991) 'Cold Fusion II: the story continues' **New Scientist** 129 No. 1752 (19th January 1991) pp. 50–53

Bockris J. O'M. and Veziroglu T. N. (1991) **Solar-Hydrogen Energy** Macdonald Optima, London

BP (British Petroleum) (1991) **BP Statistical Review of World Energy** (annual)

Briggs D. and P. Smithson (1985) **Fundamentals of Physical Geography** Unwin Hyman, London

Brown G. C. and E. Skipsey (eds.) (1986) **Energy Resources: Geology, Supply and Demand** Open University Press, Milton Keynes

BWEA (British Wind Energy Association) (1982) **Wind Energy for the Eighties** Peter Peregrine, Stevenage

Chapman J. D. (1989) **Geography and Energy** Longman, Harlow

Close F. (1991) 'Cold Fusion I: the discovery that never was' **New Scientist** 129 No. 1752 (19th January 1991) pp. 46–50

Coggle J. E. (1983) **Biological Effects of Radiation** Taylor and Francis, London

Cook E. (1971) 'The Flow of Energy in an Industrial Society' **Scientific American** 225/3 (September) pp. 135–144

Cotgrove S. (1982) **Catastrophe or Cornucopia** John Wiley, Chichester

Crabbe D. and R. McBride (1978) **The World Energy Book** Kogan Page, London

DEn (Department of Energy) (1975) **UK Oil Shales: Past and Possible Future Exploitation** Energy Paper No. 1 HMSO, London

DEn (1979) **Combined Heat and Electrical Power Generation in the UK** Energy Paper No. 35 HMSO, London

DEn (1981) **Tidal Power from the Severn Estuary** Vols. I & II Energy Paper No. 46 HMSO, London

DEn (1984) **CHP/DH Feasibility Programme Stage 1** Energy Paper No. 53 HMSO, London

DEn (1988) **Prospects for the Use of Advanced Coal Based Power Generation Plant in the UK** HMSO, London

DEn (1989) **The Severn Barrage Project – General Report 1989** Energy Paper No. 57 HMSO, London

DEn (1990a) **Digest of United Kingdom Energy Statistics (1990)** HMSO, London (annual)

DEn (1990b) **Energy Use and Energy Efficiency in the UK Domestic Sector up to the Year 2010** HMSO, London

Dorf R. C. (1981) **The Energy Factbook** McGraw-Hill, New York

Dunn P. D. (1978) **Appropriate Technology** Macmillan, Basingstoke

Eckholm E., G. Foley, G. Barnard and L. Timberlake (1984) **Fuelwood: The Energy Crisis That Won't Go Away** Earthscan, London

ERL (Environmental Resources Ltd) (1983) **Acid Rain: A Review of the Phenomenon in the EEC and Europe** Graham and Trotman (for GEC), London

ETSU (Energy Technology and Support Unit) (1990) **The Severn Barrage Project – Detailed Report (5 vols.)** ETSU (TID 4060) Harwell

Flood M. (1983) **Solar Prospects: The Potential for Renewable Energy** Wildwood House, London

Foley G., P. Moss, and L. Timberlake (1984) **Stoves and Trees** Earthscan, London

Foley G. (1987) **The Energy Question** Penguin Books, London

Foley G., P. Moss and L. Timberlake (1984) **Stoves and Trees** Earthscan (IIED), London

Greenhalgh G. (1980) **The Necessity for Nuclear Power** Graham and Trotman, London

Gregory D. P. (1973) 'The Hydrogen Economy' **Scientific American** 228 (January) pp. 13–21

Gribbin J. and M. Kelly (1989) **Winds of Change** Hodder and Stoughton, Sevenoaks

Hunt S. E. (1980) **Fission, Fusion and the Energy Crisis** Pergamon Press, Oxford

IEA (International Energy Agency) (1987) **Renewable Sources of Energy** OECD, Paris

IEA (1988) **Emission Controls in Electricity Generation and Industry** OECD, Paris

IEA (1989a) **Electricity End-Use Efficiency** OECD, Paris

IEA (1989b) **Energy and the Environment: Policy Overview** OECD, Paris

IEA (1989c) **Energy Policies and Programmes of IEA Countries: 1988 Review** OECD, Paris

IPCC (Intergovernmental Panel on Climate Change) (1990) **Climate Change** Cambridge University Press

Kemp D. D. (1990) **Global Environmental Issues: A Climatological Approach** Routledge, London

Kenward M. (1991) 'Fusion becomes a hot bet for the future' **New Scientist** (16th November 1991) p.10

Kirwan D. F. (ed.) (1987) **Energy Resources in Science Education** Pergamon Press, Oxford

Kovarik B. (1982) **Fuel Alcohol** Earthscan (IIED), London

Kovarik T., C. Pipher and J. Hurst (1979) **Wind Energy** Prism Press, Dorchester

Krause F., W. Bach and J. Kooney (1990) **Energy Policy in the Greenhouse** Earthscan, London

Leggett J. (ed.) (1990) **Global Warming: The Greenpeace Report** Oxford University Press

McCormick J. (1985) **Acid Earth: The Global Threat of Acid Pollution** Earthscan (IIED), London

McMullan J. T., R. Morgan and R. B. Murray (1983) **Energy Resources** Edward Arnold, London

McVeigh J. C. (1983) **Sun Power** Pergamon Press, Oxford

McVeigh J. C. (1984) **Energy around the World** Pergamon Press, Oxford

Marjoram T. (1982) 'Energy pipe dreams in the Pacific' **New Scientist** 12th August 1982 pp. 435–438

Miller G. Tyler (1990) **Resource Conservation and Management** Wadsworth, California

Munslow B. with Y. Katerere, A. Ferf and P. O'Keefe (1988) **The Fuelwood Trap: A Study of the SADCC Region** Earthscan, London

Mustoe J. (1984) **An Atlas of Renewable Energy Resources** John Wiley, Chichester

NRPB (National Radiological Protection Board) (1989) **Living With Radiation** HMSO, London

NEA (Nuclear Energy Agency) (1989) **Nuclear Energy in Perspective** OECD, Paris

Odell P. (1989) **Draining the World of Energy** in R. J. Johnston and P. J. Taylor (eds.) **A World in Crisis?** pp. 79–100 Basil Blackwell, Oxford

OECD (1988) **Environmental Impacts of Renewable Energy** OECD, Paris

Open University (1984) Audiovisual material for Block 5 of Course S238: **The Earth's Physical Resources** Open University Press, Milton Keynes

Park C. C. (1987) **Acid Rain: Rhetoric and reality** Methuen, London

Patterson W. C. (1990) **The Energy Alternative** Boxtree, London

Pentreath R. J. (1980) **Nuclear Power, Man and the Environment** Taylor and Francis, London

Popper K. (1972) **Objective Knowledge: An Evolutionary Approach** p. 298 Oxford University Press

Porritt J. (1990a) **Friends of the Earth Handbook** Macdonald, London

Porritt J. (1990b) **Where on Earth are we Going?** BBC Books, London

Porteous A. (1991) **Dictionary of Environmental Science and Technology** Open University Press, Milton Keynes

Raistrick A. and C. E. Marshall (1939) **The Nature and Origin of Coal and Coal Seams** English Universities Press, London

Ramage J. (1983) **Energy: A Guidebook** Oxford University Press

Reay D. A. and D. B. A. McMichael (1988) **Heat Pumps** Pergamon Press, Oxford

Rees J. (1985) **Natural Resources: Allocation, Economics and Policy** Methuen, London

RCEP (Royal Commission on Environmental Pollution) (1976) **Sixth Report** HMSO, London

RCEP (1986) **Eleventh Report** HMSO, London

Schumacher D. (1985) **Energy: Crisis or Opportunity?** Macmillan, Basingstoke

Seager J. (ed.) (1990) **The State of the Earth** Unwin Hyman, London

Seneviratne G. (1991) 'Money Pledged for International Fusion Project' **New Scientist** 23rd November 1991 No. 1796 p. 18

Slesser M. (1988) **Macmillan Dictionary of Energy** Macmillan, Basingstoke

Slesser M. and C. Lewis (1979) **Biological Energy Resources** Pergamon Press, Oxford

Sørensen B. (1979) **Renewable Energy** Academic Press, New York

Soussan J. (1988) **Primary Resources and Energy in the Third World** Routledge, London

Sverdrup H. U., M. W. Johnson and R. H. Fleming (1942) **The Oceans, their Physics, Chemistry and General Biology** Prentice-Hall, Englewood Cliffs, New Jersey

Taylor R. H. (1983) **Alternative Energy Sources** Adam Hilger, Bristol

Twidell J. W. and A. D. Weir (1986) **Renewable Energy Resources** E. & F. N. Spon, London

WCE (Watt Committee on Energy) (1979) **Energy from the Biomass** Report No. 5 The Watt Committee on Energy, London

WCE (1981) **Assessment of Energy Resources** Report No. 9 The Watt Committee on Energy, London

WCE (1984a) **Nuclear Energy: A Professional Assessment** Report No. 13 The Watt Committee on Energy, London

WCE (1984b) **Acid Rain** Report No. 14 The Watt Committee on Energy, London

WCE (1985) **Small Scale Hydropower** Report No. 15 The Watt Committee on Energy, London

WCE (1989) **The Chernobyl Accident and its Implications for the UK** The Watt Committee on Energy, London

WCED (World Commission on Environment and Development) (1987) **Energy 2000: A Global Strategy for Sustainable Development** Zed Books, London

WEC (World Energy Conference) (1989) **Survey of Energy Resources** WEC, London (triennial)

White L. P. and L. G. Plashett (1981) **Biomass as a Fuel** Academic Press, New York

Wild M. (ed.) (1980) **Energy in the '80s** Longman, London

Wilson E. M., C. S. Haine and R. H. Hamer (1980) **Small Scale Hydroelectric Potential of Wales** Chadwick-Healey, Cambridge

Wilson J. I. B. (1979) **Solar Energy** Taylor and Francis, London

World Bank (1991) **World Development Report (annual)** Oxford University Press

Most of the organizations concerned with energy supply and distribution are very pleased to give assistance to the serious inquirer.

For conventional sources these may include the many oil companies operating in the North Sea, British Gas, British Coal, NIREX, National Power, Powergen, Scottish Hydroelectric, Scottish Power, National Grid; and for renewable sources the Department of Energy's Energy Technology and Support Unit (ETSU) at Harwell, Oxfordshire, and the Centre for Alternative Technology near Machynlleth, Powys.

The supply and use of energy is also an important concern of organizations such as the Association for the Conservation of Energy, Friends of the Earth and Greenpeace.

For keeping up to date with developments in the energy field, there are several specialist journals, but *New Scientist* and the quality press are perhaps the most accessible sources. All the TV channels transmit relevant programmes from time to time.